Solid Waste Management

Editors
Velma I. Grover
B.K. Guha
William Hogland
Stuart G. McRae

A.A. BALKEMA/ROTTERDAM/BROOKFIELD/2000

ISBN 90 5410 786 3

A.A. Balkema, P.O. Box 1675, 3000 BR Rotterdam, Netherlands
Fax: +31.10.4135947; E-mail: balkema@balkema.nl
Internet site: http://www.balkema.nl

Distributed in USA and Canada by
A.A. Balkema Publishers, 2252 Ridge Road, Brookfield, Vermont 05036, USA
Fax: 802.276.3837; E-mail: Info@ashgate.com

Foreword

Solid Waste Management is from many perspectives an international mission. Pollutants and emissions do not know of any borders, they move from country to country causing problems. We are all involved in the important task to protect the environment from these emissions—emissions coming from different products which finally have turned into waste. Wastes are simply discarded products and the design of a product can have a very significant impact on the nature of the waste that is produced. Different products are spread all over the world, finally ending up as waste, thus contributing to the negative impact on the environment if not handled in a proper way. It is only by working together from an international perspective that we can develop ways of handling produced waste in such a way that the trans-boundary movement of emissions and the negative impact on the environment is minimized.

The relationship between correct and good solid waste management and the protection of human health, natural resources and the environment requires a continuously upgrading solid waste management. In many places in the world today solid waste management starts with a controlled sanitary landfilling, developing into a more advanced system with source separation of hazardous products and recyclable products and a more advanced combination for the recovery and final disposal of waste. It is necessary to prevent requirements within the solid waste field to promote high environmental practices, and in such a way also continuously upgrading solid waste management. One way of upgrading solid waste management and promoting high environmental practices is by presenting good examples and models of how well functioning solid waste management in different countries results in the protection of human health, natural resources and the environment. An increased international exchange of models, examples and experiences will contribute to a faster development of a correct and good solid waste management.

Solid waste management in practice must be based on integrated systems with a combination of many different methods. There should not be any contradiction between the different methods—instead they should be regarded as complementing each other. With increasing demands of the saving of non-renewable natural resources, increased recycling, an even better environmental protection and an upgraded solid waste management emphasis could be placed on the following principles for the future handling of waste:

- Recovery and recycling of materials, involving direct action by residents and industries. The conservation of resources should take place mainly within the industry by waste minimization and by material recovery and recycling.
- An increased and improved collection, transportation, recycling, treatment and final disposal of hazardous waste.
- Thermal treatment of waste, with energy recovery—for heat, cooling and/or power production.
- Utilization of the easy biodegradable fraction of organic waste for composting or anaerobic digestion with gas production.
- Landfilling, which always will be needed regardless of other methods utilized, for non-recyclable materials, residues from other treatments etc.

The choice of methods or combination of methods must be based upon a number of conditions—the market for different recycled products, the possibility of using produced energy and compost, the area available for landfilling, the population density etc. It is not possible from an International Perspective to give an overall recommendation or solution on the waste problem. The decision must be based upon the local, regional and natural conditions.

Hakan Rylander
President ISWA 1996-98

Preface

Solid Waste Management has emerged as one of the important areas of overall urban policy planning and management. The UN Conference on the Environment and Development held in Rio, in 1992, affirmed that environmentally sound management of waste is one of the major issues that need to be tackled for maintaining the quality of the earth's environment and to achieve environmentally sound and sustainable development. In Agenda 21, a separate Chapter, chapter 21 entitled "Environmentally Sound Management of Solid Waste and Sewerage Related Issues" is there to handle this problem. This acts as a guideline for the national agenda of the countries, which will vary depending upon the local conditions and needs.

The papers in this book demonstrate that Solid Waste Management is not only the problem of the developing nations but also of the developed nations, though the nature of problems of both differ. The issues related to waste management could be traced back to a few factors—inability of the local authorities to pick up all the waste, lack of proper data, lack of financial resources, lack of required skills, improper disposal facilities, either lack of legislation or the outdated legislation, and improper organizational structure of the authority responsible for waste management. Another important issue is the attitude and concerns of the people.

The studies show that waste management should be looked as an integrated problem instead of looking at the technical (collection, transportation, disposal), social, economic and financial factors separately. Therefore, for proper waste management a number of factors—storage, collection, transportation, recovery, reuse, recycling, minimization, treatment and disposal—have to be taken into account. Specially, in the developing countries there is a need to address the issues of better equipment, better health and working conditions for the workers, a routine for collection and transportation and to adopt scientific treatment and disposal methods. Besides this, the authority must take into account the climate, cultural and social factors, standard of living, level of social and economic development, the skills available and technology suitable for the local conditions. For example, as pointed out by one of the authors in Chapter 1, because of the cheap labour in the developing countries, it might be cheaper to go in for the manual composting rather than the mechanical treatment. A note of caution is necessary here, as prevention is better than cure—it is not always advisable to go in for the cheapest option (though at times the cheaper option is a better option—like the example cited above).

Another common theme is the need for strong legislation. To discharge the functions efficiently, there is a need for proper legislation, organization and the will to implement such laws. Therefore, the national or local government should enact the necessary laws and should continually update it.

Another problem of deep concern is the problem of scavengers. It could be related to poverty—they take this job out of sheer necessity to survive. Unlike the claims made, the scavengers do this job to earn their living and not for the purpose of recycling or for the preservation of environment. They collect the material which is well paying and ignore the recyclable which does not fetch them enough money even if the material is hazardous to the environment. Therefore, attention must be paid to their quality of life and the inhuman treatment they go through. Although, the existence of *kabariwala* or hawkers (who, buy things from households or institutions) to collect waste can be traced back, the presence of scavengers (who pick up things from the thrown waste) is new because of the growing poverty. So, the scavengers call for immediate political and social attention while the hawkers do not. On the one hand, while the existence of the recycling industry in the developing countries is traditional, the recycling industry in developed countries can be attributed to the need to recycle waste because of concern for the environment and limited availability of natural resources.

Finally, the chapters of this book have also reflected a need for education and public information and public participation. The purpose of the education is twofolds. First, to influence the behavioural attitude and concepts of ethics and morality of individuals as well as communities, so that they understand the need and importance of managing the waste for improving public health and the quality of the environment. Second, to train skilled professionals for sound waste management. As the behavioural attitude and concepts of ethics and morality of individuals are developed in childhood, it is necessary to impart the knowledge of the solid waste as a part of the general subject of environment (which will view the problem in a holistic manner rather than as an isolated problem).

It can be safely concluded that every country needs to develop its own solid waste management policy, which is best for its economic, social and environmental needs. Moreover, there should be active people's participation at every stage of waste management, close public-private-non-governmental organization partnership and there should be stress on training all the key actors involved in waste management.

Velma I. Grover

Contents

SECTION 4: COUNTRY CASE STUDIES OF EUROPEAN CONTINENT

SECTION 5: COUNTRY CASE STUDIES OF LATIN AMERICA

Introduction

Until recent times, changes in the environment have been almost entirely the product of natural forces. However, over the last two centuries human activity has begun to play an increasingly significant role in bringing about the changes in the environment, largely due to two developments: the industrial revolution and unprecedented growth in human population.

The industrial revolution applied new forms of energy and organization to economic activity, thus multiplying human productivity many times and giving human beings greater power to transform the environment than ever before. The next global revolution—the Information Technology revolution—has increased the mobility, and the faster telecommunication and the computers have made this earth a small place to live in. The impression of the accelerated industrialization; increased and mechanized agriculture; economic globalization and cultural exchange; technological innovations; and new political and world order can be seen everywhere, in that the nature is also not left unaffected.

In fact, because of these revolutions, the influence of human activity on the environment is at a scale that rivals the forces of nature. However, the human activities—the principal forces behind the change today—are the forces whose control lies directly in our hands.

By the late 1960s or 1970s people in the industrialized nations began to recognize the interdependence of man and the environment, and became increasingly concerned about pollution. As human society started feeling the brunt of its activities, it took some steps towards the prevention in the degradation of the environment—the first such step was the Stockholm Conference in 1972. The next giant step in this direction was the Rio Conference in 1992, which concluded with the fact that if the human society has to endure, not just for another century but for thousands and thousands of years, we need to learn a way of life that can be sustained. Along with the major problems like climate change, ozone depletion and loss of biodiversity, it also recognized waste management as one of the problems that is threatening the survival of mankind (Agenda 21). Although, waste management may not have major impact at the global level, yet waste mismanagement can spread epidemics like plague and dengue. So, human society must learn to develop more efficient technologies that produce as little harmful waste as possible and must learn to rely on resources that are renewable.

For proper waste management it is necessary to understand the existing system of waste management and problems associated with it. There are many books written on Solid Waste Management—so this book does not look at what is solid waste management and the general methods of collection, transportation and disposal. Solid waste is the garbage produced by households. Solid waste management is the management of this waste—it consists of three broad steps—collection, transportation and processing or disposal. Collection, generally, includes collecting waste from the source of production (like households, markets, schools, institutions etc.) and depositing it at a place from where the waste is picked up by the responsible authority (like bins in developing countries and outside the houses for curbside collection in developed nations). The waste is transported to the processing or disposal site. Processing consists of either resource recovery, waste recycling or composting of organic waste. Disposal is generally the dumping of waste at landfill sites (which can be engineered or uncontrolled sites) or incineration of waste (at places, where energy is recovered from the incinerators. This is not strictly considered a disposal method—it is energy recovery).

There was no problem of waste management during the ice age or when man settled down from being a hunter to a farmer; the waste which was produced was either used in farming or was assimilated by nature. The needs were less and the resources available were plenty. The problem started mainly after industrialization when the needs of the resources grew tremendously, waste was produced in a quantity which could not be assimilated by nature, the chemicals produced and disposed cannot be recycled by nature. Now, concerted effort is needed to restore the socio-ecological balance of nature for sustainable growth.

This collection of papers written by international experts in an effort to highlight present waste management practices and their shortcomings. The first section is devoted to general waste management, followed by sections on Africa, Asia, Europe, and Latin America.

The lead article, in the first section by Drs Pieter van Beukering and Joyeeta Gupta, **"Integrated Solid Waste Management in Developing Countries—a Concept of Many Dimensions"**, discusses the waste hierarchy followed by a discussion, which brings out differences between the waste conditions in the developed and developing countries and its impacts on waste hierarchy and waste management. It further brings out the importance of the informal waste recovery sector and the implications of domestic and international trade of recyclable materials. The waste and gender issues follow this. The next paper by Dr. William Hogland and Ms Marcia Marques, **"Waste Management in Developing Countries"**, reviews the waste management situation in developing countries and gives an overview of the problems and possible solutions. After discussing the problems related with urbanization, the paper looks at the issues of sorting, collection, transport, recycling and disposal methods. It also brings up the issue of hazardous waste, environmental and health aspects of waste mismanagement and importance of education (especially for the young generation) to improve the solid waste management systems. Issues of maintenance and institutional setup follow this.

With the growth in population and urbanization, municipalities face the problem of providing the basic necessities. One of the ways to address this sort of the problem is privatization. Faith S. Oro, in **"Privatization, Partnership and Participation: The Challenge for Community Involvement in Municipal Solid Waste Management"**, gives a literature review on privatization and institutional aspects of solid waste management. The author looks at the role of public and private sectors, barriers to developing public-private arrangements and the challenges for community participation. **"The Enforcement Issues for Environmental Legislation in Developing Countries: A Case Study of India"** by Dr. D.C. Pande and Velma I. Grover gives first an account of the environmental laws in India. The major environmental concerns of India were focussed on water pollution, air pollution, soil erosion, deforestation, desertification and loss of wildlife. It shows that the purpose of environmental regulation is to prevent, mitigate or remedy environmental damage. Even though, there are plenty of laws to deal directly with environmental issues, there are hardly any laws to deal with solid waste problems, and the ones that exist are outdated.

The second section explores a number of important issues in solid waste management, facing Africa. The **"Introduction to Waste Management in African Continent"** by Santosh Kumar, R. Bappoo and M.B. Sasula, gives a general overview of the problems of waste management in some African cities like Lusaka, Nairobi, Pretoria and Gaborone. Chapter six, **"Solid Waste Management in East and Southern Africa: the Challenges and Opportunities"** by Dr. Nonita T. Yap, gives a review of the current situation of waste management in eastern and southern Africa. It discusses the present situation of waste generation, characteristics, collection, recycling and disposal, thus bringing out the problems in this part of the world drawing on examples from South Africa, Uganda, Cameroon, Ghana, and Kenya among other places. It then points out that laws do exist for waste management in the African continent but some of the policies are outdated. It further highlights the importance of NGOs in the collection and recycling of waste, especially in South Africa and the leading and active role played by women in this sector, the paper also discusses the emerging trends in solid waste management in that part of the world. It concludes with the note that the problem is not intractable but tremendous financial resources will be needed for this. Chapter seven, **"Solid Waste**

Management: A Developing Country's Perspective" by S. Kumar, R. Bappoo and M. B. Sasula gives information, procedures and suggestions in the implementation of solid waste management in developing countries, giving examples from Zimbabwe. According to the authors, very few countries in Africa have really addressed themselves to the issues related to the environment. Instead most of them have been concentrating on border disputes or wars, internal conflicts, ethnic violence, political power which have rendered their economies weak, and thus have very little funds set aside for environmental issues. Hence, there has been very little or no interaction on solid waste management and its complexities. They conclude with recommendations for improvement in a solid waste management system in the city of Bulawayo, Zimbabwe. Chapter eight, **"Deployment of vehicles from a Central Depot to Collection Sites when Each Vehicle Makes at the Most One Trip"**, by S. Kumar and R. Bappoo gives the allocation of refuse vehicles used in the city of Bulawayo, Zimbabwe, from the central depot to a number of collection sites.

Section three is devoted to Asia. This section begins with the **"Introduction to Waste Management in Asia"** and looks at the problems in various aspects of solid waste management like data collection and planning, collection, transport, disposal and processing of waste in the Asian continent. This also looks at the costs (per capita expenditure on Solid Waste Management) in different Asian countries. The chapter, **"Solid Waste Management of Hong Kong: Its Evolution and What Developing Countries can Draw On"** by Chung Shan Shan discusses the waste management system in Hong Kong and illustrates the mistakes to be avoided by the policymakers in developing countries. Starting with the evolution of waste management system in Hong Kong in 1997 it gives the definition of the municipal solid waste in Hong Kong and the reasons for the decline in the generation of municipal solid waste in the island. It gives waste reduction, waste recovery, scavengers, waste trade and difficulties of waste recovery sector in the city. Waste disposal methods and lessons drawn for developing countries from experiences of Hong Kong follow this. In Chapter eleven, **"Sustainable Waste Management in Metro Manila, Philippines"**, Dr. Leonarda N. Camacho, gives an account of the present Solid Waste Management practices in the city followed by a discussion on the changes required in the metro. The next four chapters describe waste management in four cities of India. The chapter on **"Collection, Transportation and Disposal of Municipal Solid Wastes in Delhi—A Case Study"** by Dr. D.K. Biswas, S.P. Chakrabarti, and A.B. Akolkar discuss the present situation and problems of solid waste management in Delhi. This is given in light of the Supreme Court case (Union of India vs. Wadhera) giving details of the duties bestowed upon the Central Pollution Control Board to look into the problems of solid waste management in the city and to address public grievances. The paper, **"Public and Private Partnership in the Urban Infrastructures"**, by Dr. Pradeep Monga gives an account of the present status of waste management in Shimla bringing out the problem areas, innovative approaches adopted to regulate and manage the problem of solid waste in the city. It highlights the importance of people's participation for better solid waste management and also covers the plans and strategies developed for effective waste management in the city, the public-NGO and public-private partnership, and the projects started in the city for the processing of waste in an organized and scientific way. The next chapter, **"Urban Waste Management in Hilly Towns"**, by Dr. Virinder Sharma, offers another perspective of Himachal in particular, Shimla. He looks at the recycling aspect of waste management giving the role of various voluntary agencies in Himachal Pradesh (H.P.)—a State within the Republic of India. It details various initiatives taken by the secretariat of Himachal Pradesh, State Council of Shimla and NGOs to address the mounting problem of waste management in H.P. The last chapter in the Indian section, **"The Management of Municipal Solid Waste Using Flexible Systems Approach"**, by Dr. Sushil discusses waste management problems followed by various processing and disposal methods. The main emphasis of the paper is the development of a methodology utilizing flexible systems approach with the aid of system dynamics and physical system theory in order to manage Municipal Solid Waste in an effective and logical framework. The author has discussed the different variables required and the computer simulation package used for this programme. This model has a wide applicability for local planners and can be utilized for source segregated, centrally segregated or nonsegregated waste.

"Solid Waste Management in Sri Lanka: A Country Perspective" by Dr. Ajith de Alwis gives a brief introduction to Sri Lanka, its administrative structure and environmental problems. Then in a description on solid waste management, he discusses solid waste generation and management, the various disposal facilities in the country followed by resource recovery discussing in detail about scavengers. He states that there is no proper established method of solid waste management - it is collected and disposed off at a large number of unprotected sites. He also notes that, although most of the waste is organic and can be processed, this is not done. He concludes the chapter by discussing the future of waste management in Sri Lanka.

The next section focuses on Europe and starts with **"The Introduction to Waste Management in Europe"** which gives a brief introduction to the waste management practices in Estonia, Lithuania, Latvia, Sweden, the UK, Germany, Hungary, Spain, Italy and Yugoslavia. The stress is more on the disposal practices. It gives a comparison of the cost of waste disposal on landfill sites and incineration in various countries and the amount of waste incinerated with how much percentage is incinerated with energy recovery. The future legislations for the disposal of municipal solid waste in the continent is also discussed. Dr. Stuart G. McRae, in **"Waste to Energy—the UK Situation"**, describes the various types and amounts of waste generated annually in UK and UK waste management strategies. The reasons for dominance of landfill is followed by an account of various disposal methods—landfill, recycling and incinerator, highlighting the advantages and disadvantages of each. Dr. Joan Mata Alvarez, in **"Environmentally Compatible Strategies for the Integrated Management of Municipal Solid Wastes"** discusses the strategies to handle waste—ways in which society can change the manner in which it handles waste, giving example of Spain. The first part of the paper addresses the reduction and recycling strategies while the second part deals with technical innovations like anaerobic digestion. The next paper, **"Forestry, Municipal and Agricultural Wastes for Fuels and Chemicals Production in Yugoslavia—Resources and Engineering Data"**, by Todorovic M.S., T.M. Stevanovic-Janezic, F. Kosi, M.G. Kuburovic, G. Koldzic and A. Jovovic, provide an overview of waste disposal and wood residues for energy and chemical production in Yugoslavia and Serbia. Authors also discuss the recycling of residues from essential oil production into forest ecosystem. This is followed by the quantity and composition of municipal solid waste, disposal methods—landfills, low oxidation methods, anaerobic digestion—practiced in the country concluding it with mathematical modelling of bioreactors.

The last section is devoted to Latin America. It begins with the introduction **"Solid Waste Management in Latin America and the Caribbean"** by Dr. William Hogland and Ms Marcia Marques which addresses the main aspects of solid waste management in Latin America and the Caribbean comparing it with that in the developed countries. The waste composition, characteristics, collection, transportation and treatment options in use in the region are discussed. It also discusses the participation of different multilateral financial agencies and aid organizations in the improvement of waste management systems in the region during recent years. The last chapter of the book, **"Solid Waste as Ore: Scavenging in Developing Countries"** by Dr. William Hogland and Ms Marcia Marques examines some aspects of the social and productive aspects of the scavenging aspects in developing countries. The authors also point out the similarities and differences between two communities of scavengers at landfill sites in Rio de Janeiro State, Brazil. It also discusses the social forces driving the organization of the scavengers and their hierarchical structures.

Velma I. Grover

Section 1

The Issues

1

Integrated Solid Waste Management in Developing Countries

Pieter van Beukering and Joyeeta Gupta

Institute for Environmental Studies (IVM), Vrije Universiteit
Boelelaan 1115, 1081 HV Amsterdam, the Netherlands

INTRODUCTION

Looking at historical citations one can assume that integrated solid waste management is a concept which gradually developed throughout time. For example, in many European countries in the 1660s, burial in cotton or linen shrouds was banned to allow more cloth for paper making (World Resource Foundation, 1997). In 1896, the first combined waste incineration and electricity scheme began operation in east London. Until 1894, New York's garbage was mainly dumped in the Atlantic Ocean, polluting the beaches, resulting in protests by the resorts on the shores of New Jersey and New York. With the inauguration of the new administration, a "radical program of source separation was implemented on the premise that mixed refuse limited the options for disposal, whereas the separation of wastes at the source allowed the city to recover some of the collection costs through the resale and reprocessing of materials" (Gandhy, 1994: 72). In the early 20th century, an ethnic minority in Egypt—the Zabbaleen—was one of the world's first communities to integrate recovery and recycling of municipal waste (Baaijens, 1994: 10).

Yet, it was the environmental movement in the late 1960s which formally presented integrated solid waste management as a guiding principle for sustainable handling of societies' residues. Since this was also an economically prosperous period in most industrialized countries, waste managers were not pressured by narrow budgets to implement these ideas. Thus, until the economic recessions of the mid-1980s the new paradigm of integrated solid waste management was widely implemented in industrialized countries (Schall, 1995). Integrated solid waste management is a very broad concept. Essentially it implies that decisions on waste handling should take into account economic, environmental, social and institutional aspects. The integration can take place at various levels: (1) the use of a range of different collection and treatment options, (2) the involvement and participation of all the stakeholders and (3) the interactions between the waste system and other relevant systems such as industry (Lardinios and Klundert, 1997).

In theory this concept seems rather attainable. In practice, however, it proves difficult to include all these aspects at the same time. While integration is already strenuous in industrialized countries, it is even more difficult in the developing countries which suffer from limited municipal budgets for waste management. As a result, policies tend to focus mainly at the waste hierarchy. This paper will argue that since the social, economic and institutional factors in developing countries are quite different from those in the industrialized countries, for a waste management policy to be successful, it needs to adapt concepts developed in the North to suit the circumstances in the South.

This paper first explains the concept of the waste hierarchy. This is followed by an analysis of the differences in the waste conditions and management between industrialized and developing countries. This has certain implications for the application of the waste hierarchy. The importance of the recognition of the informal waste recovery sector in integrated solid waste management is explained in the following section. The dilemma of optimal waste management within administrative units is discussed along with the implications for trade. Following an analysis of the multi-disciplinary approach within integrated solid waste management and the importance of gender in waste management conclusions are drawn.

THE WASTE HIERARCHY

Within integrated solid waste management, the waste management hierarchy has been taken as a key element in waste management policy. Especially in the North, the hierarchy is widely applied as a guiding principle. The hierarchy is based on environmental principles, and implies that waste, depending on its characteristics, should be handled by different methods: a certain amount should be prevented by either reducing the content of waste or by reusing the waste, another share of the waste stream needs to be converted into secondary raw materials, some parts can be composted or used as source of energy, and the remaining may be landfilled. Especially in developing countries, reality does not adhere to this sequence. In fact, a large quantity of waste is dumped in an uncontrolled manner, or worse, burned in the open air. Obviously, as shown in Fig. 1, these options do not belong to the waste hierarchy because of their unacceptable high levels of environmental damage.

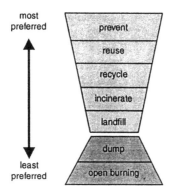

Fig. 1. The waste hierarchy

Although, this ranking of waste management options does provide policy makers with an effective basis, it should be realized that integrated solid waste management goes beyond the waste hierarchy. It is generally known that the hierarchy has to be applied in a flexible way and it is only meant as a guideline to achieve the best environmental solution in the long term. Still, the hierarchy has always been subject to fierce criticism. For example, many believe that the options presented in the hierarchy should not be ranked in a particular order but considered as a "menu" of alternatives. "It is not a question of good and bad waste management options or technologies. Rather, each option was equally appropriate under the right set of conditions addressing the right set of waste stream components" (Schall, 1995). In determining whether the hierarchy is applicable in developing countries, the following section evaluates the essential differences between the North and the South.

NORTH AND SOUTH

Research indicates that there are significant differences between the industrialized and the developing countries that could be of relevance to solid waste management strategy. It should be realized however, that even within these individual categories of countries, notable variations occur. The composition of waste and the managerial problems vary from country to country and from city to city. Differences in demography and climate make Jakarta and Mexico City as distinct as Tokyo and Helsinki. Differences in classes within countries also implies that there is a difference in wastes depending on the income of the people. The rich not only produce larger quantities of waste but their waste may have a higher value.

There is one factor which has a decisive impact on solid waste: wealth. Waste generation is closely related to income. The level of waste management service is also indirectly dependent on the available budget. Moreover, this relationship is reflected in the composition and the quality of the generated waste flows. The waste in developing countries is generally characterized by a large proportion of organic waste, moisture content and ash. The solid waste stream in industrialized countries is generally dominated by a large amount of inorganic materials. However, this also largely depends on the level and type of recycling. Figure 2 shows that in the mid-1980s, Japan clearly did not maintain a well-developed recycling system for organic waste.

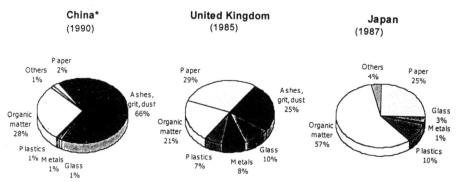

Fig. 2. Composition of national waste flows (Source: UNEP 1996)

Another factor that influences waste management is the sociocultural factor. While in many parts of Asia organic waste has been traditionally seen as a resource and has been used as animal feed, to fertilize the lands, and as a source of fuel for cooking; in Africa waste is regarded as dirty and people are generally averse to close contact with it. Religion and culture may also have an influence on the way society deals with waste; in Islamic cultures people are generally averse to any close handling of waste; in Hindu cultures although waste is seen as a resource, activities around waste management tend to be undertaken by the lower castes (Lardinois and Klundert, 1993: 20).

Since the characteristics of the waste accumulated are different and since there are financial and institutional bottlenecks, the problems that waste managers face in developing countries are also quite different. Their problems tend to be more in relation to imperfect collection, inadequate receptacles, poor transport, health hazards and poor disposal practices. In addition, waste management in developing countries is highly labor intensive. It is therefore not surprising that the type of waste management is different between the North and the South (see Fig. 3). As a result of the institutional bottlenecks and the labor intensity of the collection processes, there are severe health hazards to both those directly concerned and those who live in the neighborhood.

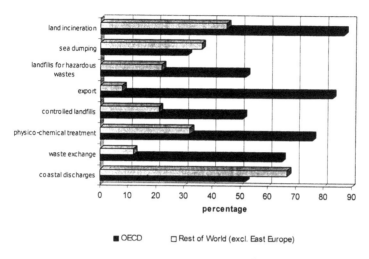

■ OECD　　□ Rest of World (excl. East Europe)

Fig. 3. Percentage of countries reporting availability of different types of waste treatment options and disposal facilities. *Source: UNEP 1996*

The waste management hierarchy is especially adopted and disputed in the North. Only a limited number of studies indicate the relevance of the hierarchy in the South. A paper presented during a recent session of the International Solid Waste Association (ISWA) reported that the traditional hierarchy should not be emphasized for the management of municipal solid waste under low-economic conditions (Ogawa, 1996). Unfortunately no clear reasons were provided in the paper.

Two modifications between the desired waste hierarchy of the North and the South may be required. First, the ranking of options may differ. For example, in the South incineration may be considered a worse option than landfilling given the high moisture content of the waste. Second, the importance of certain options will deviate between industrialized and developing countries. In the overview below, a reasoning of potential differences for each option in the hierarchy is listed:

- The potential for *waste minimization* is smaller in developing countries because of the generally low level of consumption and the low degree of the luxury of consumption. Significant reduction in consumption cannot be expected of people who just live at subsistence levels. Having said that, many developing countries have reached the take-off point in their development process such as the newly industrializing countries of Southeast Asia and many countries have a large well-developed industrial sector such as India, China and Brazil and in these countries waste minimization at source may be a significant policy tool to follow so as to avoid large waste streams in the future. In northern countries, a gradual switch is taking place in the emphasis from control to prevention as a waste management strategy, although this is moving quite slowly. Consumers are being motivated to reduce their waste generation through waste fees by weight or volume. Producers are being invited to prepare a life-cycle analysis of their products and to take a more responsible approach to management.

- The potential for *reuse* is larger in developing countries because consumer preferences are less specific about the fashion value of product, and more about their functional merits. Moreover, cheap labor enables a high degree of repairing activities which would be uneconomical in industrialized countries. Having said that, it must be acknowledged that almost everything that can be reused is reused in these countries.

- The potential for *mechanical recycling* is larger in developing countries because of the cheap labor. Recycling is a production activity which utilizes a relatively simple technology which can be operated on a small scale and which is labor intensive. Closely related to this option is the ability to sort waste into different groups which also enhances the recyclability.

- The potential for *chemical recycling* is on the one hand smaller because of the high development costs of this relatively new technology. On the other hand, this option may be feasible in developing countries because of the low cost of manual sorting.

- The potential for *thermal recovery* is smaller because of the high moisture content of the municipal waste caused by the large extent of organic materials in the solid waste stream. This would enhance the feasibility of *composting* which indeed is becoming more and more popular in many developing countries.

- Recovery and recycling markets are growing in both the North and the South. In the North this trend mainly evolves around developing recycled content products. The changes in developing countries focus on the improvement of the quality of secondary materials while simultaneously developing new markets. International supply sources for secondary materials are increasingly utilized. Furthermore, the trend in intermediate processing is significantly different. In the North, intermediate processing focuses more on capital intensive methods by the development of mechanical and electro-mechanical sorting and separation techniques which reduce manual labor.

- The potential for *landfilling* is smaller because of the lack of funds to develop controlled alternatives and the increasing lack of space in the rapidly utilizing mega-cities. There are also health risks associated with landfills and the risk of fire. The sites for landfills need to be carefully located and monitored. In the North there is a continuous emphasis to minimize the volume which is being sent to landfills. With only modest efforts this volume is expected to reduce by 5-20%. A variant of landfills can be named the landfill farming variant. In Calcutta, about 20,000 people lease filled landfills from the municipality and grow 25 varieties of vegetables, yielding on average 150-300 tons per day (Lardinois and Klundert, 1993: 19). This potential (also referred to as urban farming by some) is quite exhaustive but can only be safe under the condition that no toxic waste or other comparable pollutant reaches the landfills;

- The potential for *incineration* is low since the waste tends to have a high organic content; and is more moist, thus rendering it less suitable for incineration (Lardinois and Klundert, 1993: 17). While incineration has the plus points of producing energy, reducing the volume of the waste relatively quickly to a relatively small volume that is also generally quite sterile and therefore unlikely to lead to infectious sicknesses, it tends to be very capital intensive, requiring specialized man-power and can lead to toxic emissions (Buekens and Patrick, 1985: 129-130).

The above conceptual analysis does not yet reflect the practical problems of implementing the waste management hierarchy in developing societies. For a good waste management system, it is necessary to have an extensive overview of the waste streams. Measurement systems have gradually become well developed in the North and are presently a prerequisite of proper management. For example, managers are now aware of the impact of a successful recycling program on the composition of the remaining waste flow so that incineration installations can be adjusted. In most developing countries, proper characterization of waste is absent. As a result, management based on technology designed for northern waste is imported, even though it may not be suitable in the local context (Savage and Diaz, 1995).

Another prerequisite for a good waste management system is an administrative structure that is coherent and conducive to good management. In many developing countries (and some industrialized countries), since the waste management options often lie not within the jurisdiction of one policy maker, efficient and consistent implementation is rather difficult. For example, in some countries tariffs of the lowest ranked options are lower than the ones of the higher ranked ones. Also, in some countries, private companies are paid per ton of waste collected, which does not stimulate waste prevention and recycling (Lardinois and Klundert, 1997). Schall (1992) writes in his conclusions that the waste hierarchy "must be understood from both a solid waste system and a production system perspective. To do this, we must think about solid waste within a framework that includes production level issues: decisions about what to produce, how much to produce, and what to use in terms of raw material inputs into those production processes." This requires a clear and coherent administrative structure.

A good waste management system also needs to take into account the climate in these countries. It is very well to ask residents to separate the waste, but high temperatures may mean that such waste needs to be picked up with greater regularity. The climate may also have implications on the nature of the landfill and on the type of recycling into organic waste. Nevertheless, although there is no readily applicable script for waste managers in developing countries, the hierarchy may form an inspiring concept for them to look beyond the conventional options practiced.

FORMAL AND INFORMAL WASTE SECTORS

At present in developing countries, up to 50% of municipal budgets are spent on solid waste management. Due to lack of funds, on an average only 50% of the generated solid waste is collected in their cities (Cointreau-Levine, 1989). Even when budgets are adequate for collection, safe disposal remains a problem. The limited resources for Solid Waste Management (SWM) by municipalities in developing countries enhance the need for cost-effective options to manage urban solid waste. Traditionally, waste management was a responsibility of the government and was, therefore, commonly performed by municipal officials. Yet, integrated waste management also demands the cohesion of present and potential actors within waste management. Two tendencies can be identified in developing countries which could promote this integration.

First, a general trend of privatization can be recognized in the waste sector. In the city of Manila, the local government has contracted out 100% of the waste collection to one private company to cover the entire city. In nearby Quezon City, the local government has divided the city in six franchise areas each serviced by one contractor. In Malaysia, collection and disposal of municipal solid waste in the entire country will be fully privatized in the coming years. The country has been divided into four zones to be serviced by four SWM consortia. The government and the private firms are now in the process of negotiations and are mapping out the technical requirements and performance indicators. Privatization is also the trend in Korea, Singapore and Indonesia (Lapid, 1997).

Whether this trend really improves waste management in the South remains to be proved. Evidence from OECD counties is that private collection probably reduces costs by 10-40% but the evidence for developing countries is unclear (Pearce and Turner, 1994). The introduction of private collections confirm significant cost reductions in Sao Paolo relative to Rio de Janeiro (Bartone *et al.*, 1991). In general, there appears to be scope for private operators. The examples in the Philippines and Malaysia show, however that the privatization trends in developing countries are of a monopolistic nature. Such a monopolistic trend is unlikely to be efficient because of the lack of competition. Nor can it be justified by the economies of scale argument since economies of scale appear not to exist in collection and disposal beyond a certain point.

The second phenomenon in the South is the existence of a large informal sector which operates in parallel with the formal waste collection authorities. This informal sector is mainly guided by market forces. As a result it seems logical that the informal sector is more efficient than the formal sector. Their role in urban centers is essential for both environmental and economic reasons. By collecting waste materials the informal sector takes over a part of the burden of the municipalities. Since the waste recovery is labor intensive and involves the skills that are already abundantly available in the community, it provides a livelihood for many new immigrants and marginalized people in metropolises. Estimates show that these activities account for an estimated 1-2% of the workforce in large cities (Cointreau-Levine, 1989). In Madras, the number of waste pickers were estimated to be about 30,000 which is approximately 1% of the population (Sudhir *et al.*, 1996). More than half of the waste is collected by the informal workers in Cairo (Lardinois and Klundert, 1993: 14). Moreover, the informal collection reduces capacity problems at dumpsites and promotes recycling.

It is difficult to quantify the total contribution of the informal sector to urban waste management. The informal nature of this sector inherently implies lack of official statistical data. Quantification of informal recovery is therefore scarce and uncertain. For Mexico, waste pickers are estimated to remove 10% of the municipal waste (Bartone *et al.*, 1991). In Bangalore (India), the informal sector is claimed to prevent 15% of the municipal waste going to the dumpsite (Baud and Schenk, 1994). In Karachi, the informal sector reduces municipal waste collection by 10% (Ali *et al.*, 1993).

The actors of the informal sector should be incorporated into the formal sector and for instance be provided with sanitary working conditions (Ogawa, 1996). Yet, such recognition quite rarely occurs. In Indonesia, there is presently a draft decree on solid waste management under discussion among the concerned national agencies which among other things promotes the encouragement of waste pickers and buyers by the local government and/or the private contractors (Lapid, 1997). Municipalities in developing countries are slowly beginning to recognize the merits of the informal recovery sector. As a result some policy makers try to accommodate scavengers in their policies.

Yet, generally waste pickers are accused of being an obstacle to the operation of solid waste collection, recycling and disposal services. In Karachi, the municipal authorities are considering ways to formalize recovery by way of introducing closed containers, establishing transfer-cum-sorting stations etc. This may disturb the delicate balance of the existing set-up and threaten the vast efforts at recycling which are being undertaken by the informal sector (Ali *et al.*, 1993). On the other hand, it is feasible that their activities can be effectively incorporated into a waste recycling system. Such an opportunistic approach is required for sustainable development of solid waste management programmes in developing countries. Clustering of informal activities creates synergy and increases the knowledge base. Also, the reservation of specific locations for recovery activities would improve the performance of the informal sector. At present, investments in such locations remain low because of the high risk of recycling firms to be forced to leave the often illegal locations.

Another effective tool would be to upgrade the overall image of informal and formal waste workers. The social status and income of these management workers is generally low especially in developing countries. This owes much to a negative perception of people regarding the work which involves the handling of waste or unwanted materials. Such people's perception leads to the disrespect for the work and in turn produces low working ethics of laborers and poor quality work. The lack of public awareness and school education about the importance of proper solid waste management for health and well-being of people severely restricts the use of community-based approaches in developing countries (Ogawa, 1996).

DOMESTIC AND INTERNATIONAL SYSTEMS

The term "recycling" has two dimensions—recovery and utilization. Recovery refers to the diversion and collection of waste materials from landfills, incinerators, or other disposal methods. Utilization refers to the processing of diverted waste into new and useful materials and products. In recent years the industrialized countries of the North have observed significant increases in the quantity of waste recovered and utilized. These trends have resulted from higher disposal costs, increased public concern about the health and environmental impacts of waste disposal, and a general perception that recycling can result in resource conservation. In many northern countries, policies have been adopted to encourage or mandate the recovery of waste materials. Policies have also been adopted to mandate the utilization of wastes—for example, mandated recycled material content in selected products and government procurement practices that favor recycled materials.

Another trend is the increasing trade of secondary materials between the North and the South, but also between countries in the South. Recovered waste materials are being increasingly exported to the South for utilization. As a result, the North has developed into a net supplier of recyclable waste while the South has developed into a net importer. As is the case with any commodity, intentional trade of secondary materials allows countries with different comparative advantages to exercise those advantages to bring about a more efficient allocation of resources. In the absence of market failures, international trade in secondary materials allows economic gains in both the North and the South (Beukering and Curlee, 1998). Several cases have been reported in the literature which support this positive relation between trade and recycling:

- Namibia does not posses proper facilities to recycle waste sufficiently since the quantities produced domestically are very small. Export channels might be a reason to collect and store these wastes until the quantity is sufficient to export it to appropriate countries. Therefore, different materials are collected throughout the country and transported to South Africa where they are recycled. Presently it is not feasible to set up recycling plants in Namibia due to the small amounts of recyclable materials available. Lack of adequate water is a hampering factor. Recycling of waste from Namibia in South Africa takes place for glass, cans, used oil (Kohrs, 1996);
- It has been estimated that the waste recycling industry in Colombia provides employment for 1 to 2% of the labor force. International trade in recyclable materials takes advantage of the existing differences in technical capabilities and the need for raw materials. For many years, Colombia has been importing scrap iron from the Netherlands which serves as an important input in the industry (Pacheco, 1992).
- About 90% of the waste paper collected in Hong Kong is exported. Unlike waste paper, all aluminium cans are exported since no company producing aluminium cans operates in Hong Kong (Yeung and Ness 1993). Without the possibility of exports this waste would have to be disposed or incinerated.
- "In Phnom Penh, garbage collection trucks are rarely seen on many streets but the city's, mostly-female, waste pickers are a common site. These pickers go from door to door to pick-up reusable or recyclable items. The materials are then sold to middlemen who export most of these items to Vietnam and Thailand" (Lapid, 1997).
- Several waste materials are collected by itinerant waste buyers and waste pickers in Kathmandu. After sorting and cleaning these materials are exported to the neighboring country India where these materials are recycled. Again, it is doubtful whether recovery would be performed without the demand from the Indian recycling industry (Beukering and Badrinath, 1995).

However, when market failures occur—such as when the health and environmental effects are externalized—international trade may lead to an increase rather than a decrease in total environmental damage. Further, international trade in secondary materials may lead to development patterns in the South that are in contrast to the preferences of both the South and the North. The "leak" of recovered materials abroad may reduce the incentive to set up recycling facilities, domestically. Moreover, relatively cheap imported materials may damage the local market for recyclable waste. Therefore, some waste experts claim that policies to promote re-use, recycling and minimization of waste generation should include measures to protect the local recycling market against the importation of cheap waste materials from the industrialized countries (Klundert, 1997).

Whether international trade in secondary materials has positive or negative environmental, economic and social impacts, can only be determined case by case, and country by country. The authors also believe that closed loop cycles are the ultimate goal. Yet, trying to solve all problems within local or national borders is not rational. Waste management should not conflict with international trade which enables a more efficient allocation of materials. In other words, integrated solid waste management should also encompass a broad "international" dimension. Not only are these issues relevant for national policy makers who must decide about legislation concerning this type of trade; these issues also are important to international interest groups, such as the Basel Convention on the Control of Transboundary Movements of Hazardous Wastes and the World Trade Organization (WTO).

WASTE AND GENDER

Traditionally waste management, like many other management issues, has taken a gender neutral approach. Such an approach has serious limitations when it comes to the successful implementation of management options, since many of the assumptions underlying such an approach are not valid. A gender approach implies taking into account that the relationships between men and women and the roles they play in society are social constructs and not the result of genetic differences. As such, these relationships tend to be context driven and can be modified over time. Such assumptions can be summed up as follows:
- the issues that concern men are similar to the issues that concern women; and
- the nuclear family is the appropriate analytical unit at the family level.

Such an approach ignores the special needs and problems of women, not only in relation to themselves, but also in their relation to men and children in society. Most women in developing countries are in the poorest group in their society; they work very long hours everyday and have three kinds of tasks: survival tasks, household tasks and income generation tasks (Dankelman and Davidson, 1989a: 3-6).

There are several roles and relationships that women have with waste. As consumers of products available in the market and in the fields—they determine the type of household waste that is generated. As mothers and carers the impacts of waste in the neighborhood on the health and well-being of the family are important for them. As people employed in the recycling/reuse or other waste-related sector, they can be seriously affected by the quality and quantity of the waste. In some societies, women are involved in rag picking, collecting waste from the streets; while the men have the tasks of collecting recyclables and selling them to recycling units. The introduction of modern technologies to facilitate the waste management process tends to displace women in their role in this sector (Lardinois and Klundert, 1993: 21). As those implementing waste policy in the domestic context, i.e., as those who will be responsible for the crucial tasks of separation of wastes their convenience and methods need to be taken into account. Finally as decision-makers in society they have a role to play. Women are also seen as key agents of change because they have specialized knowledge on resource management, they are able to work effectively together, as mothers they have a powerful influence on changing attitudes towards the environment and empowering women will also result in improving the local environment (Dankelman and Davidson, 1989b: 171-189). This is not just theory, but there are several examples of activities undertaken by women in the local management of waste (Lardinois and Klundert 1993: 28-31) provide examples of successful examples of educated women running environmental hygiene awareness campaigns along with waste collection in Bamako, Mali; of successful experimental resource recovery centers run by women in Kampal, Uganda, the Balikatan Women's Movement in Philippines, etc.

CONCLUSIONS

This paper has argued that the circumstances in the developing countries are vastly different than those in the industrialized countries. This implies that notions in relation to waste management need to be revised to make them applicable to developing countries in general and to the particular context in which it is to be implemented. Waste management strategy for the South could build on three connected strategies:

1. Identify the structural features characterizing current waste management in society and try to make it more efficient within the domestic constraints.
2. Identify the strong points and try to build on them.
3. Identify the weak points and try to eliminate them.

Identify structural features: The major structural features in the South can be generalized as follows. The existence of a large informal sector with context relevant specific tasks undertaken by women, men and children that are dependent on this work for their livelihood. Furthermore, the administrative structures are likely to remain somewhat complicated in relation to the establishment of an integrated management system. The aim should be thus to make optimum use of the informal sector and to provide it with the institutional support and help it needs to develop into an informed, safe and efficient sector. The fragmented administrative system calls for a discussion and rational allocation of responsibilities in relation to the waste management sector.

Identify strong points: Given the large discrepancies between developing countries it is difficult to give a single common strong point which counts for all countries within this group. The major strong point in the Hindu/ Asian culture is their tendency to see waste as a resource; and to be naturally inclined to conserve and reuse this resource. There is a tendency for this cultural feature to be eroded by the fast pace of development amongst the relatively wealthy community in this group. Other societies will also have their strong points. Another major strong point in these countries is the high labor intensity of the waste management system. These points should be optimized as far as possible.

To achieve this, the tendency towards resource conservation should be given new directions. Separation of waste at source is cheapest, environmentally sound and an easy first step in waste management. Public awareness campaigns combined with innovative collection services making use of the existing informal and formal sector should ensure that societies, continue to, and increase the separation of the waste at the source. The task is likely to fall on the women in the household and on men in industry. Thus, any system focusing on promoting waste separation at source should focus on the specific target group. In order to motivate these groups into feeling inclined to separate the waste, there should be clear instructions as to the need for, the relevance of, the logistics of waste collection systems and the associated rewards and penalties (economic, health, social). The involvement of the local people in developing ideas of how the waste can best be separated and collected and what incentives and knowledge would lead to a change of behavior is vital. Thus women in households in India may separate the glass and paper waste in the hope of being able to sell them to the itinerant waste buyers.

Further separation of the inorganic waste from the organic waste is desirable in order to facilitate the reuse of the organic and the inorganic waste. This implies that the traditional tendency to separate the waste at source to sell to the informal collectors should be nurtured even though the nature of consumption and waste changes in the urban households. Thus, for example, special attention needs to be paid to issues ranging from special types of waste such as plastic waste to items in the waste packet that can cause injuries to the waste pickers. To insure regular collection of the recyclable and disposable waste, community participation is crucial. The most successful initiatives in waste separation and collection in developing cities generally find their roots at the neighborhood level.

To the extent possible, substituting labor by technology would increase the energy intensity of the waste management system, would create unemployment in labor rich countries, would increase the financial burden on the municipality, and is more likely to negatively affect the women labor force. Thus for example, manual composting costs is about $1-5 per ton of organic material and about US$ 11 per ton using mechanical treatment in India (Lardinois and Klundert, 1993: 38). In such a situation, perhaps unthinkable in a northern country where labor costs are very high, manual composting methods remain the most economical. This does not however mean that the health and labor conditions of those working in this sector does not need to be improved. But replacing the labor by technology would be unwise, because of the high associated costs and the attendant unemployment.

Cheap labor does not only create economic possibilities within the countries' borders. Differences in the endowment of recyclable waste can be effectively used through international trade of secondary materials. From the perspective of developing countries, two points of view can be taken. On the one hand, if the waste generated is of very small quantities it would be unwise to invest in a domestic recycling facility since the economies of scale may not be achievable. In that case it may be more efficient to export the recovered materials. On the other hand, if a country possesses a domestic recycling industry, imports of secondary materials often form a useful supplement to the domestically supplied materials. Such an approach can be even more beneficial if it appears that the quality of the output of the recycling industry can be improved as a result of quality secondary materials (Beukering and Duraiappah, 1997). In other words, policy decisions with regard to waste management need to look beyond national borders.

Identify weak points: The weaknesses of these countries can be divided in different categories. First, the most urgent weakness is the imperfect collection, transportation and storage. Second, a relatively high degree of injuries and waste-related diseases occur among waste workers. Thirdly, the monopolistic tendency in the privatization operations can be considered a weak point in many developing countries. Such concentrated power structures threaten the efficient allocation of rewards related to waste management. Resources should be spent on ensuring that the collection services are improved; that the receptacles for receiving waste are more hygienic and more safe and that the transport system is economical and rational. If the collection systems are improved there will be less health risks to those at risk. Finally, the administrative structures should try and ensure that the privatization of the enterprise does not lead to a new, and by definition, inefficient monopoly, nor should such privatization trends displace the existing informal sector. Such privatization trends should be encouraged in those sectors where there are, at present, gaps in the management process.

Integrated solid waste management forms a very valuable basis for the improvement of waste system in developing countries. In the paper, an attempt was made to demonstrate that developing countries should develop their own concept of integrated solid waste management. Direct adoption of the northern version is doomed to fail. It was demonstrated that integration goes beyond the waste hierarchy. Other dimensions such as gender, internationalization, North-South relations, and the co-existence of a formal and an informal system, should also be taken into account in a strategy of integrated solid waste management. In that respect, both the North and the South still have a long way to go, towards creating a more sustainable system of waste management.

REFERENCES

Ali, S.M., Cotton, A. and Coad, A. 1993. Informal sector waste recycling. Paper presented at the 19th WEDC Conference on Water, Sanitation, Environment and Development. Accra, Ghana pp. 153-155

Baaijens, A. 1994. In Cario kom je als vuilnisman ter wereld. *Internationale Samenwerking.* No. 11, pp. 10-11.

Bartone, C., Leith, L., Triche, T. and Schertenleid, R. 1991. Private sector participation in municipal solid waste service: experiences in Latin America. In: Waste Management and Research 9, pp. 495-509.

Baud, I. and Schenk, H. (eds.) 1994. Solid Waste Management: Modes, assessments, appraisals and linkages in Bangalore, Manohar. New Delhi.

Beukering, P.J.H. van and T.R. Curlee. 1998. Recycling of materials: local or global? In: Vellinga, P., J. Gupta and F. Berkhout (eds.): *Managing a Material World.* Kluwer Academic Press, Dordrecht. 229-239.

Beukering, P.J.H. van, and Duraiappah, A. 1997. International Trade and Recycling in Developing Countries: Waste Paper in India. *Warmer Bulletin.* No. 52. pp. 8-9.

Beukering, P.J.H. van, and Badrinath, G.D. 1995. Challenges for the recycling industry in Nepal. *Warmer Bulletin.* No. 46. pp. 6-7.

Buekens, A. and Patrick, P.K. 1985. Incineration in WHO (ed.) Solid Waste Management: Selected Topics, World Health Organization, pp. 129-130.

Cointreau-Levine, S.J. 1989. Solid waste recycling: case studies in developing countries. In: Regional Lapid, D. 1997. *Southeast Asia regional consultation report: summary of findings.* Center for Advanced Philippine Studies. pp. 1-4.

Dankelman, I. and Davidson, J. 1989a. Why women, in Dankelman, I. and Davidson, J. (Eds.) Women and the Environment in the Third World: Alliance for the Future, Earthscan Publications, pp. 3-6.

Dankelman, I. and Davidson, J. 1989b. Working together for the Future, in Dankelman, I. and Davidson, J. (Eds.) Women and the Environment in the Third World: Alliance for the Future, Earthscan Publications, pp. 171-180.

Gandhy, M. 1994. Recycling and the politics of urban waste. Earthscan Publications Ltd., London.

Klundert, A. van de 1997. *Policy aspects of urban waste management.* Issue Paper for the Programme Policy Meeting Urban Waste Expertise Programme. April 1997. pp. l-8.

Kohrs, B. 1996. Waste management in Namibia. *Warmer Bulletin.* No. 50. pp. 10-11.

Lapid, D. 1997. *Southeast Asia regional consultation report: summary of findings.* Center for Advanced Philippine Studies. pp. l-4.

Lardinois, I. and A. van de Klundert 1997. *Integrated sustainable waste management.* Paper for the Programme Policy Meeting Urban Waste Expertise Programme. April 1997. pp. 1-6

Lardinois, I. and A. van de Klundert 1993. Organic Waste: Options for small-scale resource recovery, Urban Solid Waste Series 1, WASTE/TOOL.

Muller, M.S. 1996. "Gender and waste management". pp. 9-13

Ogawa, H. 1996; *Sustainable solid waste management in developing countries.* WHO Western Pacific Regional Environmental Health Centre (EHC). Paper presented at the 7th ISWA International Congress. Kuala Lumpur. pp. l-8.

Pacheco, M. 1992. Recycling in Bogota: developing a culture for urban sustainability. *Environment and Urbanization,* Vol. 4, No. 2. pp. 74-79.

Pearce, D. and Turner, R.K. 1994. Economics and solid waste management in the developing world. CSERGE Working Paper WM 95-05. London.

Savage, G.M. and Diaz, L.F. 1995. Future trends in solid waste management. In: International Directory of Solid Waste Management 1995/96; The ISWA Yearbook. London. pp. 22-28.

Schall, J. 1995. Does the Solid Waste Hierarchy Make Sense? A technical, economic and environmental justification for the priority of source reduction and recycling. PSWP Working Paper # 5. Yale University. New Haven.

Sudhir, V., Muraleedharan, V.R. and Srinivasan, G. 1996. Integrated solid waste management in urban India: a critical operational research framework. *Socio-Economic Planning Science.* Vol. 30, No. 3. pp. 163-181.

UNEP 1996. *Environmental Data Report 1991/1992.* Third Edition. United National Environmental Programme. Oxford, pp. 336-359.

World Resource Foundation 1997. Information sheet: paper making and recycling. *Warmer Bulletin* No. 55.

Yeung & Ness 1993. Three recycling industries in Hong Kong: market structures, vulnerabilities and environmental benefits. *Asian Journal of Environmental Management.* Vol. 1, No. l, pp. 51-59.

2

Waste Management in Developing Countries

William Hogland[1] and Marcia Marques[2]

[1] Department of Technology, University of Kalmar, P.O. Box 905, SE-39129 Kalmar, Sweden

[2] Department of Environmental Engineering, Rio de Janeiro State University, Rio de Janeiro, Brazil

INTRODUCTION

Today, the world's population generates enormous amounts of waste. Whether we live in industrialized countries (ICs) or developing countries (DCs), in big or small cities, or whether we are rich or poor, we all produce waste. For decades, ICs have lived beyond their resources. Yesterday's belief that all resources were eternal has now changed. Waste minimization is a key concept in ICs. The amount of waste produced is dependent on the country, type of urban district, population, city size, culture, style of life and, of course, income.

However, the question remains, how should waste be handled safely and economically? It may be easy to say that people should minimize the total amount of packing and municipal solid waste (MSW) in ICs where the waste production per person is very high but in DCs where generation per person already is low there is not much scope for reduction. In certain ordinary districts in the USA, each person generates 2.7 kg MSW daily, whereas in Pakistan approximately 0.5 kg waste per person, per day is generated. Some African countries generate less than that (as low as 0.2 kg per person, per day), i.e. only 10% of the amount produced in the USA on a per capita basis. However, in many DCs there is a large difference between the consumption patterns of the rich and those of the poor. Based on this difference, people in high and medium income areas can produce waste of the same magnitude as the average value for the industrialized world.

The waste management problems of the world are strongly related to growing populations and migration into cities. As a consequence, demands on natural resources, production and accumulation of waste are on the increase in such cities. The number of large cities is growing. In 1960 there were only four cities in the world with a population of more than 10 million (so-called megacities) of which only one was located in a DC. By 1980, there were eight megacities, five of which were located in DCs. The prediction for the year 2000 is 22 megacities with 17 located in DCs (see Fig. 1). Waste management will be one of the most important issues for these countries.

Number of megacities

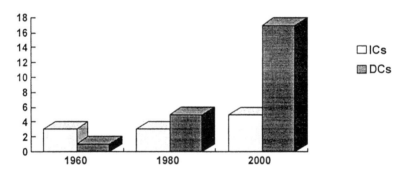

Fig. 1. Megacities with more than 10 million inhabitants, the growth of urbanization since 1960 and a prognosis for the year 2000 in industrialized countries (ICs) and developing countries (DCs)

In a city with 10 million inhabitants, a MSW production of around 5,000 tons per day and a wastewater production of 5,000-10,000 m³ per day are expected. This will lead to about 2 million tons of MSW annually and about 2–4 million m³ of wastewater. By contrast Sweden has an annual production of 2.7 million tons of MSW and 2 million m³ household wastewater (Sweden has a population of approx. 8.7 million).

Planning and designing treatment plants for amounts of solid waste and wastewater typically 10–15 times higher than those dealt with by traditional plants in ICs (mainly in Europe) is a challenge not only for local engineers but also for experienced engineers from ICs. In Fig. 2, a general picture is given of how the production of solid waste and the consumption of fresh water differ between very poor DCs, DCs undergoing economic growth and ICs.

Effective waste management is of utmost importance to a city. Collection, transport and the handling of the waste must also be properly dealt with. If not, the waste creates a number of problems, many of which are related to human health and the environment. DCs face special problems mainly

kg/person/day

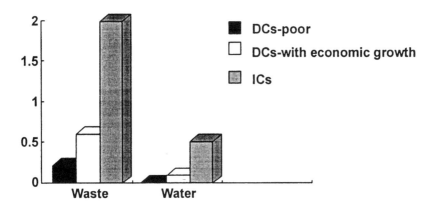

Fig. 2. Schematic diagram of the production of solid waste (kg/person per day) and consumption of fresh water (m³/person per day) in poor DCs, DCs undergoing economic growth, and in ICs

due to: (1) lack of resources for waste management; (2) low awareness and education; (3) loss of resources and energy consuming processes and; (4) problems with handling hazardous wastes. In some countries, waste is generated and handled much as in ancient times and is disposed of directly into the nearest hole in the ground outside the garden. In some cities, the manure content of the waste can also be high due to the keeping of livestock including horses, cattle, donkeys and poultry within the urban area. In big cities containers usually are not properly designed and overload is a common problem (Fig. 3).

The data and information presented here were collected by the authors when producing a technical-educational film that included three continents: Asia (Pakistan and Vietnam), Africa (Namibia) and South America (Brazil) during the period 1994-1996 (Gomes and Hogland, 1996).

SORTING

Developing countries often have an extensive sorting system. People who sort waste do so either directly on the street, at the transfer station, or on the landfill. These self-employed workers are often called scavengers. By definition, scavengers are usually people who treat waste as ore, a source from which valuable material can be extracted. For instance, in Vietnam scavengers buy recyclable goods directly from households or shops. In other countries, such as in Pakistan, scavengers acquire recyclable waste

Fig. 3. Overloaded containers for waste storage in the streets of Sao Paulo, Brazil

from the streets, in heaps outside houses or at the transfer stations, although they do not usually pay for the waste.

The existence of scavenging as a distinct occupation is based on: (1) markets for recovered materials; (2) waste in sufficient quantity and quality to meet industrial demands and (3) people who are willing or compelled to do work that is poorly paid, hazardous and of a low status.

Scavengers often live on or beside landfills in order to await the arrival of waste filled trucks. They sort the waste with their bare hands, sticks or simple hooks (Fig. 4). Sometimes entire families work at a landfill in this way.

Some youngsters work part time as scavengers to earn money for education. Some scavengers spend the productive period of their lives at the landfill while others work there temporarily when unable to find employment in the labor market. They may have ordinary jobs such as security guards, ship builders or stonemasons. In countries where the winter is harsh with temperatures dropping below zero, scavenging is rare, as scavengers' dwellings are often frugal, consisting simply of large cardboard boxes or simple shanties of wood or aluminium.

In the city of Sao Paulo (Brazil), some of the scavengers working on the streets at downtown have organized their work and created a cooperative. The COOPAMARE scavenger cooperative in Sao Paulo originated as a small group of eight scavengers in 1982. After some years and with the help

Fig. 4. Scavengers working on a landfill in Brazil

of a religious organization, they decided to create the cooperative. They obtained a grant from the Economic and Social Development National Bank (BNDES) to buy a truck, some machinery and to construct facilities such as a lunchroom and showers for the workers. The main objective of the cooperative was to increase their income and to improve the quality of their product. They sell directly to industry without go-between dealers. In the middle of the 90´s the cooperative expended 10% of the total income for the provision of water, energy, and new vehicles. The remaining 90% were shared among the scavengers according to the productivity of each. They earned around 40% more than if they were to sell to dealers and a scavenger can earn around 3 times the minimum wage of an industrial worker.

Based on the World Bank estimation, 1-2% of the population of big cities is supported directly or indirectly by the refuse generated by the upper 10-20% of the population. Scavenging not only provides a source of income to the poorer segments of the population but also reduces the need for highly sophisticated and costly recovery systems. Although there are several socioeconomic and hygienic problems associated with scavengers' activities, they nonetheless perform an important function in the recovery of resources and replace costly recovery systems. Regardless of the origin of the waste, scavengers usually sort it more finely than is common in industrialized countries. Waste is sorted into different categories including paper, cardboard, textiles, leather, aluminium cans, plastics, glass bottles, ferrous metals and wood, which is sold on to dealers. Even bone is separated out for making buttons. The quantities of paper, glass, plastics and metals, which remain in the waste stream, is relatively low in low-income countries compared to their wealthier counterparts. Bottles are also collected from shops and restaurants and are sent back to manufacturers for reuse. They can be refilled within the same country or sometimes even in other countries.

Scavengers earn their living by sorting waste and selling it on to dealers, who often live in or near the sorting area, sometimes with their entire families. The dealer carries out further sorting and then sells the materials to the industry (Gomes and Hogland, 1995).

Some municipalities own and operate fully equipped sorting plants. There are two kinds of sorting plants: (1) those that receive only recyclable waste; (2) those that receive waste that is collected without any separation at source and therefore contains a high degree of food waste.

COLLECTION, TRANSPORTATION AND TRANSFER STATIONS

Private or municipal waste management staff collects the waste. The waste is transported to the transfer stations or directly to the city dump in different ways such as: (1) being carried on the head; (2) handcarts; (3) bicycles; (4) donkeys; (5) horses; (6) ox carts and; (7) trucks (see Figs. 5 and 6).

At transfer stations (where available), the waste is then reloaded onto larger trucks for direct transportation to an incineration plant, landfill or dumping site. Tipping can be made directly on the ground and reloaded by hand or by front loader. Waste can also be tipped directly into containers and transported onward by larger trucks. About 30-60% of all the solid waste in the DCs is uncollected and less than 50% of the population is served. It is usually the poorer districts and suburbs on the outskirts of the city that are not served.

Poor and narrow roads, which are badly maintained cause damage to containers, bins, trucks and machinery. In some DCs, the trucks and machinery are imported from industrialized countries as second-hand equipment, which means that the condition of the equipment is usually poor and necessitates frequent repairs.

Several problems are related to poor maintenance and handling. In some cities, as much as 80% of the machine pool is out of order requiring repairs and spare parts. To keep the machines in good condition, spare parts must be stored and manufactured locally. Engineers specially trained to keep the machines working are also needed. Sometimes a variety of new functions for the machines must be invented. For instance, a front loader may have to load and unload trucks, spread out waste at the working face of the landfill and even perform compacting. Some larger and better

Fig. 5. Collection of paper and bottles in Hanoi, Vietnam

Fig. 6. Waste collection by bicycle in Lahore, Pakistan

equipped cities in DCs have machines and trucks for each function instead. Good maintenance is also of great importance for containers and bins. Containers are often destroyed by fires, either spontaneous or man-made. Containers are also sometimes stolen due to a lack of containers for other purposes in the country.

Transportation by oxen and horses also creates special problems: the animals must be fed and watered, the manure must be disposed of and a workshop for cart repairs is necessary.

RECYCLING INDUSTRY

Plastics are an important recyclable commodity in DCs. In Vietnam, for example, some scavengers start collecting plastics at 4 a.m. At noon they start to wash the plastics and, after drying, it is sold to a dealer (see Fig. 7). Afterwards, more plastics can be collected in the afternoon and evening, which means that these scavengers work 12–14 hours per day.

The plastic is sold by the dealers to a local plant that usually shreds it or makes plastic granules. New plastic products such as litter bags and pipes are produced in many local and simple factories (Fig. 8) although in some DCs, the technology for plastic production can be quite modern.

Paper is also a recovered commodity that ends up in the recycling industry. The biggest problems at paper mills that deal with recyclable paper are inherent impurities such as plastic, staples, stones and sand. The

Fig. 7. Scavengers washing plastic waste at a branch of Song Hong river in Hanoi, Vietnam

Fig. 8. Production of plastic bags in a small plastic recycling factory, Hanoi, Vietnam

industry must have special processes and equipment to remove these from the paper before it is processed. Products like craft paper, cardboard or toilet paper are created. Paper can be recycled seven to eight times, each time, 22 trees are saved every ton of recycled paper produced. In Brazil, 29% of paper is recycled which is comparable to the percentage recycled in many ICs.

At poor areas in Namibia, metal cans may be reused for new pots and kettles produced locally. Other metal scraps go to the metal industry. The sorting and handling of recyclable commodities in DCs represents a means of survival for many people living in big cities. In DCs scavenging brings economic benefits for the industry which obtains cheaper raw materials and eventually saves energy.

COMPOSTING

Over 60% of the municipal waste in DCs (including garden waste) is organic. The waste usually has a high moisture content, largely as a function of rainfall, especially if containers and bins are not covered. Such organic waste is suitable for composting. To increase the content of nutrients and to improve the degradation condition, the waste is shredded and mixed with manure, night soil and sludge. It is important that the sludge does not contain heavy metals or toxic substances to ensure a compost soil that is not harmful to crops. When composting waste, therefore it is always important to keep hazardous waste separate from the compostible portion. In large cities where the total production of waste is high, it might be possible to use mechanical sorting and composting plants. These hi-tech plants, if operated properly, can maintain a high standard of composting waste. In megacities, where thousands of tons of waste must be handled daily, the compost produced will be large quantity and it may be difficult to find sufficient markets for it. Waste compost has traditionally been used in DCs for soil improvement but attitudes have changed due to the risk of contaminants entering the waste stream as a consequence of industrial development.

INCINERATION

Incineration is another method of dealing with waste in some cities in DCs. But as already stated, waste in DCs contains a large amount of organic household waste with a high moisture content and low energy value (calorific value). Incineration plants do exist in the DCs but they are very few and are usually of a low technical standard leading to high emission of pollutants into the air and causing problems with the handling of ash. Incineration plants are also expensive to construct and require modern technology and qualified, skilled workers.

LANDFILLING

Dumping is the most common way of dealing with waste and landfilling is probably the cheapest way to treat waste. Therefore, up to 80% of the world practices open dumping or landfilling. For this reason, special attention should be paid to this final disposal option, mainly in DCs. In continents such as Africa, 95% of all solid waste is dumped or landfilled. Due to the lack of proper collection systems, it is still common for people in DCs to discard garbage directly onto the street, into canals and on riverbanks. As soon as dumping starts, the dumping area becomes larger. Waste attracts more waste. When the quantity of waste generated becomes a problem for the surroundings, the waste is set alight. In many cities, the smoke hangs over the roofs with a sweet, sticky smell. Such open burning of dump sites causes health problems by releasing dangerous pollutants into the air. However, the fire and the smoke kill flies and prevent the insects from breeding. In some cities, it is uncertain whether or not collected waste is taken to the city dump. Some truck drivers dump the waste along the road or, in some cases, sell it as filling material in construction areas. Using waste as filling in construction areas has caused damage to buildings because settling of it and during additional decomposition, explosive and toxic gases are generated. These gases permeate up into the buildings and represent a risk for residents.

The most common way of disposing of the waste is probably to throw it near the sea, on river banks or in rivers hoping the next high tide would transport the waste into the ocean. Some of the pollutants found in the industrial waste dissolve and get mixed with the water. If river water is used for irrigation, some of the pollutants could get incorporated into the food chain. Swimming in such polluted water should be prohibited due to health risks.

Some cities in DCs have, however, started improving the management of solid wastes. The priority has been to find a suitable location for a controlled landfill, based on hydrogeological considerations, as well as on climatic conditions. Adequate hydrogeological conditions are important to prevent groundwater contamination, mainly in those cases where financial problems impose limits on investments in bottom liners.

Even where the hydrogeological conditions have been fulfilled, the landfill design should include at least the following components: (1) a bottom layer of clay and/or a plastic liner; (2) a drainage and collection system for the leachate and surface runoff; (3) a collection system for landfill gas; (4) a treatment system for leachate that can be located on-site or in a wastewater treatment plant. The gas should either be used for energy production or flared. The closure of the landfill should include proper covering using clay and plastic liners to reduce the infiltration of surface water and to prevent erosion. A vegetation cover should also be established to reduce erosion. In order to keep the amount of incoming waste under control, a weighing bridge should be installed together with administration and personnel buildings. A fence and vegetation curtain is required to prevent people from reaching the waste and to reduce visibility. Around the landfill, groundwater wells should be built for water control.

Where the landfill is not compacted sufficiently and then covered, there is an increased risk of landslides during heavy rainfalls as it has been observed in regions with intense rainfalls.

Regarding landfill operation, in short, it can be stated that good compacting is necessary to guarantee the stability of the landfill and also reduce the volume of the waste. Good coverage also guarantees compacting and stability, while good maintenance of the horizontal and vertical drainage system is needed for leachate and gas control.

HAZARDOUS WASTE

In most waste management systems in DCs hazardous waste from industry and medical waste are mixed in the municipal solid waste stream. Waste containing chemicals such as DDT, PCBs, pesticides and heavy metals such as mercury and cadmium should be separated and not allowed to enter in the waste stream. To convince manufacturers of the need for change, they must be made aware of the fact that it is beneficial for them to maintain a high environmental profile, which is mainly dependent on the awareness of the customers.

A modern industry in a DC should be encouraged to display concern for the environment. Source separation should start close to the machine at the production site. The waste can be sorted into differently coloured bags. A special recovery yard for the separation of tyres, wood, plastic, paper, iron, ink, oil and batteries should be planned. Wastewater from factories should be treated at their own wastewater treatment plants, preferably including mechanical, biological and chemical treatment. At the very least, the sewage system should be connected to a municipal treatment facility if one exists. Unfortunately, the dominating proportion of industries in DCs represents the reverse profile and discharge directly into rivers and the sea. Solid waste from small-scale industrial enterprises is generally disposed into the municipal waste stream or directly into the surrounding environment.

There are several important aspects concerning the introduction of an industrial recycling policy in DCs: (1) the incoming materials need to be ascertained; (2) workers need to be involved in the discussions and their opinions taken into account and (3) the waste handling procedures must be subject to as much patient dissemination of information as possible. There are many advantages in starting a waste management program in a factory in a developing country. The first noticeable positive effect is the reduction of costs. Furthermore, the increased awareness among the workers filters back to their homes and helps to change the attitude and behaviour of the people.

The inadequate treatment of medical waste is also a serious problem in DCs. In most of these countries, there is a need for appropriate guidelines for the handling and safe disposal of medical waste. Although many hospitals burn their pathogenic waste, most of the existing incinerators are out of order due to the lack of fuel and/or spare parts. Consequently, pathogenic waste often ends up in the municipal solid waste stream and can cause serious health hazards, mainly due to handling by scavengers, the collection staff, and the workers at the landfill.

Some municipalities in DCs use septic ditches for the medical waste. A hole is dug out in the ground, filled with medical waste, covered with antiseptic chemicals and closed up again with soil. To use this simple method for medical waste disposal, some environmental aspects must be taken into account: the soil type at the site should have low permeability and the groundwater level should be more than three meters lower than the bottom of the ditch to avoid pollution of the groundwater. Besides the prevention of groundwater pollution, the site must be located more than 500 m away from the nearest inhabited area, to avoid scavenging. Although this technique presents several limitations, when land is available, the cost of constructing septic ditches is low in comparison with other options.

ENVIRONMENTAL AND HEALTH ASPECTS

The environmental impacts of the dumping in DCs are mainly related to: (1) destruction of local ecosystem; (2) ground and surface water contamination due to leakage and (3) pollution of the air due to the release of biogas and open burning.

Human contact with refuse implies a high risk for a variety of diseases including tetanus, typhoid, hepatitis and cholera. Infectious diseases can be spread either by direct contact with the waste, by animals such insects, birds, goats and cows, or by windborne distribution.

Scavengers as well as municipal waste management staff, usually eat their lunch amid the waste. Sometimes, scavengers collect food waste that has been dumped along with hazardous waste. It is usually not sufficient to merely place fences around the landfill even if they are electrified or to move the landfill to another site because the scavengers simply follow.

However, health care awareness is increasing, as can be seen at some sorting plants where the personnel wear masks and gloves (Fig. 9) and by the construction of facilities such as lunchrooms and showers at offices and scavengers' cooperatives.

EDUCATION

The only way to solve the waste problems of DCs is by changing the attitude of the younger generation through better education and school campaigns. Information can be disseminated at schools, on television, at special waste campaigns, and with training (particularly of waste handling staff and industrial workers). Education efforts to increase awareness must be made at all levels, from nursery schools to universities. It is necessary to intensify international contact, as well as develop international cooperation not only between DCs and ICs but also between different DCs, which are facing the same problems or facing different stages of economic development.

One strategy that has shown results in communities with low educational level among adults is to teach children. They, in turn, teach their parents. Waste management companies need special strategies for public-aided cleansing in areas with a minimal urban infrastructure. This must always

Fig. 9. Sorting plant receiving the recyclable fraction from a source collection that collects two fractions: wet waste (food) and recyclable waste (Curitiba city, Brazil).

be carried out with the participation of the community, being based on the principle that if the citizens cannot pay in cash they will still be served and may pay in kind. In order to obtain the cooperation of the locals, companies must try to implement as many small improvements as are possible. Such improvements might involve, for example, access roads, water supply or lighting and need not relate directly to urban cleansing. With such an approach they can gain the trust of the poorer communities. The programme called "Saturday is cleaning day" in Niteroi city, Brazil is an example of such approach. The municipal cleansing company has special cleaning programs for slums and low-income areas. Every Saturday between 7 a.m. and 1 p.m., people from the relevant urban district (organized in neighbourhood associations) help clean up open areas, channels and stormwater pipes. The solid waste collected is then, stored in containers that are coupled to the conventional system where tractors or trucks collect the waste.

MAINTENANCE AND TECHNOLOGY TRANSFER

There are numerous bad examples of technology transfer from ICs to DCs and high technology does not always solve problems. Every step in the waste-handling system must be studied and the potential effects analysed before introducing new systems. This must include not only the technical and economic effects but also the social effects. It is important not to try to solve all waste-handling problems at the one and the same time. Likewise, it is important to avoid large-scale constructions and plants based on technologies not tested under local conditions. Solutions need to consider the local situation and existing systems. Collection and costs need to be minimized and acceptable treatment methods based on the "minimum requirements principle" must be established. Concerning maintenance, the following aspects are of paramount importance: (1) the equipment used should be manufactured locally and designed for local needs; (2) spare parts should be manufactured and/or stored locally; (3) workshop staff should be trained by the manufacturer and (4) any machinery purchased should be of well-known brand names which are in good supply.

INSTITUTIONAL ASPECTS

In many DCs, 30% or more of the community budget is spent on waste management, usually without satisfactory results. In ICs, 1-3% of the budget is sufficient to get a satisfactory result. Lack of cost assessment and insufficient cost recovery lead to a strong dependence on external financial assistance (Pfammatter and Schertenleib, 1996). It is, for instance, expensive to let people throw waste on the street and then pay workers to pick up the same waste. The economic losses due to flooding when solid waste clogs the sewer systems or builds up under bridges must also be considered. In medium-sized and large cities, third party services seem to be preferable, mainly with regard to collection. If so, trained local government staff should be responsible for monitoring and controlling the services supplied by contractors. Adequate fees should be fairly calculated so as to ensure a sustainable system.

Every citizen, industry and commercial activity should pay according to the amount of waste generated, which is generally related to income. It is essential to improve the management and organizational capabilities of the community and to establish financially stable institutions capable of planning ahead and adapting to change.

CONCLUSIONS

Environmental problems related to the urbanization process are increasing. The main urban problems in developing countries are primarily related to the population growth and the migration of people from rural areas and smaller cities into the larger cities. Some developing countries are experiencing rapid economic and industrial growth and are gaining economic strength. However, the economic growth is not being followed by adequate measures to master the most urgent environmental problems associated with rapid urbanization and the growth of cities. The problem related to megacities located in DCs can be placed at the top of the list of well-known environmental problems of the world today. Most of these problems are water and waste-related. Infrastructure and waste treatment facilities are grossly insufficient and in many places, municipal and industrial waste is dumped in the natural environment. The present situation with respect to the pollution of air, soil and water, as well as the lack of basic sanitation facilities in these cities, create living conditions that need to be changed. If sanitation in large cities is to be improved, a integrated approach to resource management is essential. In order to achieve this, basic changes are required, not only in applied technologies but in the educational systems, policies, aid programmes, social habits, and in the structure and management of societies as well. Knowledge and responsibility should always go hand in hand, and teachers and scientists need to be involved in the sustainability process, not only as advisors but also as participants. Furthermore, the most important objective for the future is to design an adequate solid waste management system for DCs in order to improve the quality of life for billions of inhabitants living under widely varying conditions with hundreds of different cultural, historical and economic backgrounds. Can we achieve this goal using the same strategy in developing countries as in industrialized countries? The answer probably is no. However, we can blend knowledge and experience with creativity in order to find the right way.

ACKNOWLEDGEMENTS

The authors wish to thank the National Council of Development Communication (NCDC), New Delhi, India. The first author is especially grateful to the joint secretary Velma I. Grover for her merit in organizing the seminar as well as his stay in India. Thanks are also due to the Swedish International Cooperation Development Agency (SIDA) and the Swedish Board for Industrial and Technical Development (NUTEK) for their financial support.

REFERENCES

Gomes, M.M. and Hogland, W. 1995. Scavengers and landfilling in developing countries. Fifth International Landfill Symposium Caglian, Italy, 2-6 October. Proceedings Vol. I: 875-880.

Gomes, M.M. and Hogland, W. 1996. Waste Management in Developing Countries. Video Film, 33 minutes. Made in Video AB, Lund, Sweden.

Pfammatter, R. and Schertenleid. 1996. Non-governmental refuse collection in low-income urban areas: lessons learned from selected schemes in Asia, Africa and Latin-America. SANDEC, Report No. 1/96, Duebendorf, Switzerland.

3

Privatization, Partnership and Participation: The Challenge for Community Involvement in Municipal Solid Waste Management

Faith S. Oro

N.T. Yap Environmental System Analysts (YESA) Limited
Guelph, Ontario, Canada N1G 4K6

INTRODUCTION

The demand for improved approaches to solid waste management is growing. Decision-makers and practitioners in *Low Consumption Countries* (Yap, this volume) are facing the challenge of addressing the deleterious effects of the inadequate collection and disposal of solid waste on the health and productivity of both the population and the environment. The inadequacy of the provision of municipal solid waste management services solely by governments is partly due to the inability of government agencies to provide and maintain such services. Bureaucratic constraints, financial limitations and lack of relevant managerial, financial and technical skills and expertise are reasons often cited (Kessides, 1993; Ostrom *et al.*, 1993; Schübeler *et al.*, 1996). The provision of municipal solid waste collection and disposal services, however, is regarded by many as a public good (Schübeler *et al.*, 1996; van de Klundert and Landinois, 1995). As a consequence, municipalities are under pressure to respond.

Privatization, Partnership and Participation

One response gaining in popularity is the privatization of municipal solid waste management services. Privatization is commonly understood to be "any process whereby the private sector is involved in the provision of public services" (Gidman *et al.*, 1995: 13). Critics have, however, argued that the term is misleading because it connotes a minimal and subordinate role played by government agencies (Ostrom *et al.*, 1993). It also assumes that the private sector is composed only of the *formal* private sector[1] although most enterprises involved in solid waste management in *Low Consumption Countries* are classified either as part of the *informal* private sector[2], or a 'grey zone' taking on a combination of features from both sectors (van de Klundert and Landinois, 1995).

The term "private-public partnership" is used in this paper to broadly describe alternative institutional arrangements which recognize the important contributions and roles of four principal actors: the public sector, formal private sector, informal private sector and non-governmental organizations. The participation of all four actors is touted as a key to addressing recurrent problems encountered in the pursuit of an efficient, equitable, accountable, responsive and adaptable solid waste management system (Gidman *et al.*, 1995).

Organization

Based on a review of a growing body of literature on privatization and institutional aspects of solid waste management as well as the author's experience working with an NGO involved in alternative techniques to solid waste management[3], this chapter approaches the topic by providing an overview of the potential and limitations of public-private arrangements. The literature reveals that "privatization" can be, and in some cases is, more broadly and creatively defined to include not just corporate private partners but small-scale informal private partners and non-governmental organizations that often act as intermediaries. Where appropriately integrated, such partnerships between the public, formal private, informal private and non-governmental sectors can result in an improved solid waste management system which adequately and equitably addresses issues of health and sanitation, and financial and environmental sustainability. The next two sections briefly outline the varying roles played by the formal private sector, public sector, informal private sector and non-governmental organizations, and the types of collaboration undertaken by these sectors. Under the presumption that there is a danger of losing sight of the roles local communities can and do play in privatized arrangements whether formal or informal, the text concludes with a brief examination of community involvement in municipal solid waste management.

ROLES OF PUBLIC AND PRIVATE PARTNERS

Formal Private Sector

What makes a partnership with the private sector so attractive? Perhaps the strongest arguments supporting the role of private enterprises in the provision of municipal refuse collection and disposal services is that they operate more efficiently than the public sector (Ostrom *et al.*, 1993; Schübeler *et al.*, 1996) and are more directly accountable to their clients (UNEP, 1998). The private sector is specifically said to have any combination of the following competitive characteristics (Kessides, 1993):

1) the appropriate skills and expertise to manage operations of municipal SWM system;
2) an overall lower production and delivery cost;
3) being less prone to bureaucratic limitations and political influences;
4) better accountability to customers because of private enterprise's desire to achieve customer/client satisfaction;
5) financial autonomy from the government's operating expenditures and debt servicing; and
6) dynamic efficiency or "the motivation to invest in and maintain capital equipment needed to expand and introduce technological improvements" (Kessides, 1993: 21).

The advantage of the private sector is evident in Angola. Management of the decrepit para-statal company responsible for solid waste removal for Angola's capital city, Luanda, was transferred to a private company in 1997. This arrangement has resulted in noticeable improvements in the range of the area covered, as well as the frequency and regularity of the collection service provided (Development Workshop, 1998)[4].

Public Sector

Increasing acceptance of the function of the private sector in solid waste management means that the role of the public sector is shifting from one of a service provider to a regulator. The government is called upon to regulate and control the activities and monitor the performance of contracted private enterprises (Schübeler *et al.*, 1996: 33). Since continuing public interest has to be safeguarded, municipal authorities retain ultimate responsibility for the service (Gidman *et al.*, 1995: vii; van de Klundert and Landinois, 1995: 18). In Malaysia, a central complaint bureau acts as "liaison between the public, private contractors and the appropriate government agency, following up on whether the problem encountered in the provision of services resulting in the complaint has been adequately addressed" (Gidman *et al.*, 1995:14).

At the operational level, it is important that municipal solid waste authorities commit some form of managerial and financial resources to a solid waste management system. Indeed, UNEP's review of sound practices of international cases involving municipal solid waste management notes that "while private sector organizations have a role in the waste management sector, sound practice virtually requires a fiscal commitment from some level of government to design, finance, create, and maintain the municipal solid waste management system" (UNEP, 1998) and to recover some or all of the costs of collection. Outside of the public sector's managerial and regulatory functions, municipal solid waste authorities can lend credibility to small-scale operations by, for example, using such services on a contractual basis or effectively integrating these services into the formal municipal solid waste management system.

Informal Private Sector

Cases from Africa, Asia and Latin America reveal that the contribution of the informal private sector in solid waste management is being increasingly recognized by both governments and donors (Gidman *et al.*, 1995; van de Klundert and Landinois, 1995). An often cited example is the deliberate and successful integration of the traditional refuse collection system controlled by the *zabbaleen* and the *wahiya* into Cairo's formal municipal waste management system (see Assaad, 1996 for further details). Other examples from Latin American countries—such as Brazil, Colombia and Mexico and Asian countries—such as the Philippines and India, are touted whereby small-scale enterprises involved in waste collection, transfer and/ or sorting are integrated with the formal municipal solid waste collection system (UNEP, 1998).

The benefits accrued to solid waste management from the activities of the informal private sector are numerous. (1) The informal sector can recover and return to productive use materials that would otherwise end up in the waste stream. In Angola, Development Workshop mobilized community members to manually sort waste and use the separated sand and solid waste materials as infill for depressions for much needed, although temporary, road improvement (Development Workshop, 1998). (2) The informal private sector can be capable of handling large volumes of materials at little or no marginal cost to the municipal government. (3) Their waste-sorting activities for the purposes of selling, re-using and/or recycling can reduce a vast amount of waste materials requiring collection and transport. (4) Indirectly, the reduction of waste extends the lifetime of capital investments such as environmentally appropriate landfills, composting facilities and collection equipment. (5) The presence of the informal private sector in the overall solid waste management system means that formal private operators and/or municipal agencies have reduced risks: marginal activities, unpredictable costs, or unreliable revenues are transferred to the informal private operator. (6) The informal private sector are more ready and willing to provide waste removal and sanitary services to otherwise unserved and generally poor sectors in the city (van de Klundert and Landinois, 1995: 15)

Non-governmental Organizations

Non-governmental organizations (NGOs) play a key role in municipal solid waste management because as close partners of the informal sector, they offer a higher degree of credibility in the eyes of the municipality and donors. NGOs, as intermediaries between grassroots initiatives and municipal governments, the formal private sector and donors, are able to promote and advocate local concerns at a larger scale (Pfammatter and Schertenleib, 1996b; van de Klundert and Landinois, 1995), lobby more effectively than community groups and the informal sector for appropriate policy changes, facilitate and/or animate community initiatives and provide awareness-raising and skill-building support and advice to community-based groups, individual users and sometimes to the municipality and the wider public. Actual roles of NGOs in solid waste management have varied from serving as umbrella organizations under which community-level enterprises or neighbourhood committees operate, to providing a channel for donor financing[5].

TYPES OF PUBLIC-PRIVATE ARRANGEMENTS

Despite the dynamism of the informal private sector and non-governmental organizations in solid waste management, "privatized" or "public-private' arrangements are usually narrowly conceived as contractual lease arrangements between government agencies and the corporate private sector. In this misconception, the potential and actual contributions of informal private enterprises and non-governmental organizations to running an efficient, equitable and sustainable solid waste management system are overlooked.

Gidman *et al.* (1995) reveal that public-private arrangements can take a variety of management forms (Table 1). However, it is telling that they group all public-*informal* private partnerships as *cooperative* arrangements and that the role of non-governmental organizations as intermediaries is disregarded. A further exploration of "public-private partnerships" could examine: (1) the management system(s) in place, (2) the *type* and *scope* of each actor's involvement, (3) the *nature* of actors involved taking into consideration that the informal private sector is not composed of a homogenous group of enterprises, (4) the limitations and opportunities of the partnerships—including financing system, conditions for cooperative arrangements, and management and technical skills, and (5) the lessons learned from past partnership experiences.

Public-private partnerships in the delivery of refuse collection and disposal services have proved to result in reduced construction costs,

Table 1. Management system of public-private arrangements

	Management System	Description
1.	Contracting Out	Placement of contract by a public agency to an external private company.
2.	Franchising/ Concession	A private partnership takes over responsibility for operating a service and collecting charges, and possibly for funding new investments in fixed assets.
3.	Affermage	Public authority controls construction and owns the fixed assets but contracts out operations, maintenance and collection of service charges. The private company has exclusive rights to *operate* the system but bears all commercial risks (Kessides, 1993: 30)
4.	Leasing	Making use of equipment/assets without purchasing but by paying a lease.
5.	Privatization (Ownership)	Public service is entirely sold to a private partner.
6.	Management Contract	Private organization takes over responsibility for managing a service to specified standards by using staff, equipment etc. of public authority.
7.	Build, Own and Operate (BOO)	Partnership between public and private sectors whereby the private firm may build, own and operate the asset/service.
8.	Build, Operate and Transfer (BOT)	Same as BOO but the asset/service will be transferred to the public sector after a period of time.
9.	Management Buyout (MBO)	The management of well run internal functions negotiate the purchase of that function and becomes a private venture.
10.	Cooperatives	Self governing voluntary organizations designed to serve the interest of their members, working in partnership with public authorities.

increased cost recovery, increased responsiveness of solid waste management system to needs of users and, as a result, increased support and sustainability of the system (Gidman *et al.*, 1995; Schübeler *et al.*, 1996). Apart from the problems facing public-private arrangements already in operation, the development of public-private arrangements is faced with some obstacles. Some of the barriers identified include financial constraints, general institutional constraints, institutional constraints for private-public agreements, markets and technologies, and donor influences (van de Klundert and Landinois, 1995):

A. FINANCIAL CONSTRAINTS

- *Municipal Government:* No dedicated income stream for solid waste services from *Municipal Governments;* sources of financing insufficient, inadequate or out of date;

- *Formal Private Sector:* Problems with credibility; difficulty with raising capital, achieving levels of insurance required, fee collection, market guarantees and steady cash flow;
- *Informal Private Sector:* Limited or no access to financing; dependent on variable cost strategies; absence of assets and securities; considered as not being creditworthy by banks, government and/or donors.
- *Non-governmental Organizations:* Limited or no access to financing; problems with credibility; lack of organizational capacity or expertise.

B. GENERAL INSTITUTIONAL CONSTRAINTS
- *Institutional Infrastructure:* Lack of political will, appropriate intellectual framework and understanding which fosters innovative solutions; predominance of technology-oriented solutions;
- *Waste Management Personnel:* Lack of skills, expertise and interest; trained in "modern" technical solutions; prone to corruption;
- *Legislation and Regulations:* Does not facilitate the formation of cross-sectoral (public, formal private informal private and non-governmental sectors) partnerships; if legislation in place, provides few tools for coordinating or managing such partnership arrangements.

C. INSTITUTIONAL CONSTRAINTS ON AGREEMENTS WITH THE FORMAL AND INFORMAL PRIVATE SECTOR
- Public and private formal sector resistance to informal private sector involvement;
- Public sector resistance to formal private sector involvement;
- Private informal sector resistance to contracting and cross-sector partnerships;
- Barriers related to the attainment of recognition of informal sector activities and their institutionalization within the formal waste management system.

D. MARKETS AND TECHNOLOGIES
- (Partnerships involving informal sector operators depend on informal sector's ability to use the collected materials for their own manufacturing, or to prepare materials for commercial use.) Key constraints relate to marketing, or 'closing the loop' for recovered materials.

E. DONOR INFLUENCE
- Donor biases towards particular technical approaches.

- Donor interventions motivated by bureaucratic procedures (Adapted from van de Klundert and Landinois, 1995: 29-40)

While private enterprises, whether formal, informal or both, are able to provide solid waste collection, transfer and/or disposal services more efficiently and at lower cost than the public sector, *privatization does not in itself guarantee efficiency* (Ostrom *et al.*, 1993; Schübeler *et al.*, 1996). Nor will simply choosing a form of public-private arrangement result in improved service delivery. Lessons cited in the literature suggest that it is important to keep the goals of solid waste management in mind when choosing public-private arrangements. As mentioned earlier in this chapter, partnerships should aim to be economically efficient, equitable in service provision, accountable and transparent to customers and clients, responsive to varying demands and adaptable to changing needs and conditions (see Ostrom *et al.*, 1993 for further discussion).

THE CHALLENGE FOR COMMUNITY PARTICIPATION

Broadly, communities participate in solid waste management as *users* of services (such as households, families, community groups and individuals) and/or *providers* of services (such as organized refuse collectors, fee collectors and micro-enterprises). Some of the more common ways in which communities participate in solid waste management as observed in the literature are outlined below.

User Participation

User participation in solid waste management is generally limited to the primary collection of domestic refuse (Pfammatter and Schertenleib, 1996b: 18). Education and awareness-raising interventions are usually required at this level to change the perception of users regarding the importance of healthy sanitation habits and to educate users on the proper procedures for separation, collection times, location of collection points, etc. The desired outcomes of education and awareness-building initiatives include: (1) increased user cooperation and participation in improved solid waste management, (2) greater operational efficiency and coverage of service, and (3) greater willingness to pay and therefore cost recovery potential (Pfammatter and Schertenleib, 1996a, 1996b). User participation is a particularly significant issue in low-income residential areas because of the difficulty of providing incentives to more formal private operators and municipal agencies to provide refuse collection, transfer and disposal services in this area (Pfammatter and Schertenleib, 1996a; Schübeler *et al.*, 1996).

Provider Participation

Although private and community participation in municipal solid waste management is not an immediate goal of solid waste policy the participation of the private sector and community can contribute to the financial and operational efficiency and sustainability of municipal solid waste systems (Gidman *et al.*, 1995, Schübeler *et al.*, 1996). Most informal private sector activities are limited to local collection, although there have been cases of informal private sector initiatives in waste treatment (such as community composting), recovery and disposal (see Schübeler *et al.*, 1996 for further details).

Providers are not a homogenous group. They can be broadly divided between: (1) the informal sector which may or may not be organized, but which have the potential to become a private partner in primary collection of refuse, and (2) community-based organizations whose solid waste management activities may be initiated from within the community or externally organized through the use of intermediaries such as non-government organizations that are not based locally or government agencies. Examples of the former include "scavengers" and individual families involved in waste-picking, sorting and recycling. Community-based organizations include neighbourhood committees, youth and women's groups and religious organizations, all of which could be potential partners for recycling or primary collection activities (Pfammatter and Schertenleib, 1996b). In Mali, private waste collection enterprises such as Cofesfa, AGETIPE and GIE have successfully taken over collection and disposal services from the municipal agencies in some neighbourhoods (van de Klundert and Landinois, 1995).

SUMMARY

Public-private partnerships as a result of "privatization" are increasingly accepted as a means for improved municipal solid waste management. Partnership arrangements which include the public sector, the formal private sector, the informal private sector and non-governmental organizations have shown encouraging results and progress in addressing the problems of local health and sanitation, and financial and environmental sustainability. In this light, the varying roles and capacities of local communities in public-private arrangements for solid waste management require close and careful examination. Some questions to consider include: (1) How to organize and mobilize user groups and the "unorganized" informal private sector for a more efficient solid waste management system? (2) What options are available for cost-recovery and financial sustainability of a municipal solid waste management system which uses multiple service providers (e.g. formal private sector, informal private sector, public sector and non-government organizations)? (3) How best to enhance the contribution and involvement of the informal private sector?[6]. (4) What is the best way to manage "public-private partnerships" while ensuring effectiveness, efficiency and accountability?

END NOTES

[1]. Formal private sector is understood to refer to "private sector corporations, institutions, firms and individuals, operating registered and/or incorporated businesses with official business licences, an organized labor force governed by labor laws, some degree of capital investment, and generally modern technology" (Furedy in van de Klundert and Landinois, 1995: 8).

[2]. Informal private sector refers to "unregistered, unregulated, or casual activities carried out by individuals and/or family or community enterprises, that engage in value-adding activities on a small-scale with minimal capital input, using local materials and labor-intensive techniques (Furedy in van de Klundert and Landinois, 1995: 8).

[3]. From 1994 - 1997, the author worked for the non-profit organization Development Workshop (Canada) providing support to a community development and upgrading programme located in Luanda, Angola. During this time, Development Workshop (DW) began short-term emergency work in rubbish removal and drainage in the unserviced peri-urban areas of Luanda. This preliminary activity developed into the Community Based Solid Waste Pilot Project, a component of DW's Peri-urban Emergency Sanitation Project.

[4]. Peri-urban areas of Luanda are not covered in the contract arrangement. However, an agreement between the private company Urbana 2000 and local authorities of peri-urban areas allows equipment to be made available on weekends for each jurisdiction on a rotating basis (Development Workshop, 1998).

[5]. See Pfammatter and Schertenleib's (1996) detailed study examining technical, operational, organizational, managerial and financial aspects of non-governmental refuse collection in low-income urban areas.

[6]. Schübeler *et al.* (1996) argue that strengthening the organizational capacity of the informal sector is the key to effectiveness. "Support should aim to: (I) improve working conditions and facilities, ii) achieve more favourable marketing arrangements for services and scavenged materials, and iii) introduce health protection and social security measures (Schübeler *et al.*, 1996: 33).

REFERENCES

Assaad, Ragui. 1996. Formalizing the informal? The transformation of Cairo's Refuse Collection System. *Journal of Planning Education and Research* 16: 115-126.

Bartone, Carl, Janis Bernstein and Frederick Wright. 1990. *Investments in Solid Waste Management: Opportunities for Environmental Improvement.* Policy, Research and External Affairs Working Papers—Urban Development. World Bank, Washington D.C.

Development Workshop. 1998. *Community Based Solid Waste Pilot Project in Luanda's Musseques: A Case Study (Draft).* Development Workshop. Luanda, Angola.

Evan, P. 1992. *Paying the Piper An Overview of Community Financing of Water and Sanitation.* IRC International Water and Sanitation Centre. The Hague, The Netherlands.

Gidman, P. with Ian Blore, Jens Lorentzen and Paul Schuttenbelt. 1995. *Public-Private Partici-*

pation in Urban Infrastructure Services. UMP Working Paper Series 4. UNDP/UNCHS/ World Bank - UMP. Nairobi, Kenya.

Kessides, Christine. 1993. *Institutional Options for the Provision of Infrastructure*. World Bank Discussion Papers No. 212. The World Bank. Washington, D.C.

Ostrom, Elinor, Larry Schroeder and Susan Wynne. 1993. *Institutional Incentives and Sustainable Development: Infrastructure Policies in Perspective*. Westview Press Inc., Boulder, Colorado.

Pfammatter, Roger and Roland Schertenleib. 1995. Micro-enterprise:-a promising approach for improved service delivery. *SANDEC News*. No. 1, May 1995.

Pfammatter, Roger and Roland Schertenleib. 1996a. Non-governmental refuse collection systems." *SANDEC News*. No. 2, October 1996.

Pfammatter, Roger and Roland Schertenleib. 1996b. *Non-governmental Refuse Collection in Low-Income Urban Areas: Lessons Learned from Selected Schemes in Asia, Africa and Latin America*. SANDEC Report No. 1/96. SANDEC at Swiss Federal Institute for Environmental Science and Technology. Duebendorf, Switzerland.

Schertenleib, Roland and Werner Meyer. 1992a. Municipal solid waste management in developing countries: problems and issues; need for future research." *IRCWD News*. No. 26, March 1992.

Schertenleib, Roland and Werner Meyer. 1992b. Synergetic effects of municipal solid waste collection, recycling and disposal." *IRCWD News*. No. 26, March 1992.

Schübeler, Peter, Karl Wehrle and Jürg Christen. 1996. *Conceptual Framework for Municipal Solid Waste Management in Low-Income Countries*. UMP Working Paper Series 9. SKAT, Switzerland.

UMP/SDC Collaborative Group on MSWM in Low-income Countries. 1995? Cairo Declaration of Principles: Small and Micro-enterprises Involvement in Municipal Solid Waste Management Service Delivery.

UNEP. 1998. *International source boon Environmentally Sound Technologies (ESTs) for Municipal Solid Waste Management (MSWM)*. Paper published on the world wide web and available at http://www.unep.or.jp/ietc/text/TechPublications/MSWlndex.htm.

van de Klundert, Arnold and Inge Lardinois. 1995. *Community and Private (Formal and Informal) Sector Involvement in Municipal Solid Waste Management in Developing Countries*. Paper published on the world wide web and available at http://www.ias.unu.edu/ vfellow/foo/ecocity/swm-inge1.html.

4

Enforcement Issues for Environmental Legislation in Developing Countries: A Case Study of India

D.C. Pande[1] and Velma I. Grover[2]
[1] G 42 Saket, New Delhi 110 017, India
[2] 411-981 Main St West, Hamilton, On, L8S 1A8, Canada

INTRODUCTION

Leo Tolstoy wrote that every happy family is happy for the same reason, however, each unhappy family has its own reason for unhappiness. Perhaps this is valid for the nations also. Catering to the greed of vested interests seems to be one of the common causes for floundering or avoiding their responsibilities under the international environmental regime—whether it is the US in avoiding its responsibility under the Climate Change Treaty at the recent conference on the subject in Kyoto or the recent raging fire in the Indonesian forests. The most challenging problem in protection of the environment is the nation's desire to industrialize faster. This challenge is manifested in policy pronouncements of the Government of India in a similar fashion as the other nations did to meet it. Hence, the major policies are focussed on integrating the environment with developmental activities. These concerns are reflected in the national planning process, constitutional provisions and the administrative machinery set up to accommodate development policies that focus on environmental conservation. The major environmental concerns of India were focussed on water pollution, air pollution, soil erosion, deforestation, desertification and loss of wildlife. One of the unique problems with environmental laws is the issue of property rights—normally these are vested in an individual. However, elements such as air, water etc belong to the community and not an individual. So long the modern jurisprudence seeks to adore the thinnest notion of private individual interests as holding the uppermost value of the fundamental freedoms the prime issue of ecological existence would continue to be haunted with horror and hurdles in the universal pursuit of profiteering solutions to fend against the real and impending threats to the civilizational order itself.

It is generally felt that the shunning of the environmental laws in economically developing nations is due to lack of adequate legislation. In this chapter, the development of Environmental Laws in India is discussed and an attempt is made to examine the enforcement issues—whether it is due to the lack of adequate legislation or otherwise. This may not be true for all the developing nations, but is generally a good pointer.

HISTORICAL PERSPECTIVE

Conservation of forest, wildlife and natural resources has been the norm in India since time immemorial because of their link to culture and religion. Although the laws for forest conservation can be traced back to the reign of Ashcka, the forest cover in India has declined to a devastating level. As per satellite pictures, only 10% of the land has adequate forest cover, against 33% required to ensure ecological stability. India continues to lose over one million hectares of forest annually. Over 60% of India's land, some 175 million hectares is already degraded. Earlier generations used natural resources only to the extent that they could be replenished. People respected and feared nature and therefore they did not misuse it. As development continued more and more resources were used in the British era to satisfy their needs in Europe. Similarly, natural resources were used at a rate higher than they could be replenished during the industrialization period in India. Many laws have come into force since then but not necessarily implemented.

During the British era laws were very specific and dealt mainly with forests and wildlife. The laws for wildlife dealt primarily with the protection of specific areas and particular species, for example, in 1873, Madras enacted the first wildlife statute for the protection of wild elephants. After six years the Central Government enacted a broader Birds and Animals Protection Act, specifying closed hunting seasons and regulating hunting of designated species through licences. These regulations did not regulate trade in wildlife and wildlife products, and thus led to the decline of wildlife resulting in many species becaming extinct. The first comprehensive law for the protection of wildlife and its habitat can be traced to the Hailey National Park Act of 1936 under which the Hailey (now Corbett) National Park in Uttar Pradesh was established.

After independence, the rapid industrialization, desire to develop, to be self-sufficient in food, and to meet with the demands of growing population led to further deterioration of environment. Although, problems like water supply and sanitation got attention in government plans, pollution control and the preservation of environment did not get any attention till the fourth Five Year Plan (1969-1974). The subsequent, five year plans further incorporated programmes to enhance the quality of life and environment.

The year 1972 marks the watershed in the history of environmental management in India. When the 24th UN General Assembly decided to convene a conference on Human Environment in 1972 (Stockholm conference)—the Department of Science and Technology established a National Committee on Environmental Planning and Coordination (NCEPC). NCEPC is an apex advisory body for all matters concerning the protection and improvement of the environment.

However, not all laws are directly related to the issue of environment (solid waste), but have been enforced through the Penal laws which are basically meant to serve the law and order agencies. Like the Indian Penal Code, 1860 (ss. 268-269, 277-278, 290-291, 425-426, 430-432) and also ss. 134-143 of Criminal Procedure Code, 1898, The Police Act (ss. 30-32, 34)also covers the enforcement aspect with environment (solid waste) underpinning. Invasions against environment also find protection in differing methods and forms in assorted laws like Northern India Canal and Drainage Act, 1873, The Indian Explosives Act, 1884, The Indian Electricity Act, 1910, The Factories Act, 1948 and even in The Territorial Waters, Continental Shelf, Exclusive Economic Zone and other Maritime Zones Act, 1976. Most of the legislation came in force after 1972—like The Water (Prevention and Control of Pollution) Act, 1974, The Water (Prevention and Control of Pollution) Cess Act, 1977, The Air (Prevention and Control of Pollution) Act, 1981. These laws were uniformly applicable all-over India and they effectively tackled the broad environmental problems afiecting the health and safety of the citizens. All these Acts have some provisions for regulation and legal action for some specific environmental issues. However, these acts are inadequate to tackle environmental problems resulting from the fast industrialization of the country in the post-independence era. These were not only inadequate, but were found ineffective to check the degradation of the environment. In order to make law more proactive rather than reactive, in the year 1986, a comprehensive or umbrella legislation for the environment, in its entirety, was enacted by the Indian Parliament—The Environment (Protection) Act, 1986 (EPA). This relates to the protection and improvement of the environment and prevention of hazards to human beings, other living creatures, plants and property. The measures to prevent tragedies like that of Bhopal have been taken into consideration. The Act places the responsibility for laying down procedures and safeguards on the Central Government. However, Sub-Section 2 of Section 24 of EPA, 1986 makes this otherwise a very potent act, totally toothless. This particular subsection states: Where any act or omission constitutes an offense punishable under this Act and also under any other

Act then the offender found guilty of such offense shall be liable to be punished under the other Act and not under this Act. For example, the EPA Act provides for a punishment up to 100,000 rupees and imprisonment up to five years for polluting, but the Motor Vehicles Act (MVA) provides fine of Rupees 100 in the first instance and rupees 300 for subsequent offenses. According to subsection 2 of section 24 of EPA, the offender will be punished under MVA and not EPA—thus rendering EPA useless.

All these illustrations show that the various laws tend to deal directly or indirectly with environmental matters with a different set up of machineries and authorities under each of the laws. No effort has been made to coordinate the activities of one law or authority with the other to tackle the major issue of environmental pollution. It would be a worthwhile exercise to devise and deploy the resources of men and money under the shared laws with the singularity of purpose of enforcing the environmental legislation under whatever label it has identified and has been permitted to stay in the statute book. The problems can be thus summarized as follows:

a.   These enactments have emanated from various agencies with varied philosophies, technological cultures and perceptions.
b.   These do not take into account the fact that such substances can get into the life support system and also into the food chain and thus cause incalculable harm.
c.   Being punitive in nature, these come into force only after the damage is done.

CONSTITUTIONAL MEASURES

India was one the first few countries which made provisions for the protection and improvement of environment in its Constitution. In the 42nd amendment to the constitution in 1976, provisions for the protection and improvement of environment were incorporated in the Constitution of India with effect from 3rd January, 1977.

In the Directive Principles of State Policy in Chapter IV of the Constitution, Article 48-A was inserted which enjoins the State to make an endeavour for the protection of the environment and for safeguarding of forest and wildlife in the country. Another landmark provision in respect of environment was inserted by the same amendment, as one of the Fundamental Duties of every citizen of India. This is the provision in Article 51-A(g) of the Constitution. It stipulates that it shall be duty of every citizen in India to protect and improve the natural environment including forests, lakes, rivers and wildlife and to have compassion for living creatures.

ROLE OF JUDICIARY

To fulfil the aspirations laid down in the Directive Principles of the Indian Constitution, the door was wedged open (by judiciary) which expansively annotated the traditional concept of *locus standi* through liberal interpretation. Through this liberal interpretation, the underlying idea was to evolve ways and means to bring the law in line with the emerging concept of social justice—as found in the Indian Constitution.

The earlier trend has been emphatic on viewing the right through the straitjacket of an established procedure in relation to bringing legal action for violation of rights. However, the rights of the poor as is evident from the case of Municipal Council, *Ratlam vs. Vardhichand* (In Ratlam, environmental pollution affected the poor community caused by private polluters, slack and under-financed enforcement agencies and haphazard town planning. For the first time, in this case, the Supreme Court treated environment problem different from an ordinary tort or public nuisance.) and others could become possible only after liberal interpretation. The Indian judiciary has been impartial and active. The judiciary has taken pioneering steps to blaze new ways of handling new problems with old laws.

Although, the formal procedures like the Tort law have been used, a more common way of approaching the judiciary for environment cases is Public Interest Litigation. In one of the cases, a writ petition filed in the Supreme Court of India—*B.L. Wadhera vs Union of India*—the court directed the Municipal Corporation to lift garbage of all kinds on a daily basis, install incinerators, improve transhipment system for the removal of garbage, revive compost plants; appoint magisterial authorities to oversee the enforcement; and create citizen's awareness through the electronic media and other means.

An immediate identification of landfilling sites was also included amongst the directions issued. Furthermore, the Supreme Court issued directions to the statutory enacted body (CPCB) to ascertain that the collection, transportation and disposal of municipal garbage is carried out satisfactorily. The court directed the statutory agency to periodically submit to the court an inspection report to the above effect for a period of two years. The existing network of local bodies can be empowered to act in a manner that any polluter—industry or an individual—is required to remove waste or restrain itself to act in aggravating the waste through prescribed regulatory modes.

A thrust, if construed in the correct spirit, of the judgement in Dr. B.L. Wadhera's case would go a long way to streamline the management and resources for enforcing the environmental issues of dealing with the delivery, distribution and disposal of waste. Legislative machinery created under several environmental laws would remain dysfunctional in view of the lack of resources. A functional approach to the problem warrants that the agencies created under one or the other laws should be dovetailed with the network of existing units of local bodies in a manner that the resources at their disposal can be pooled and enriched together to usher a combative approach against the environmental hazards threatening our daily lives.

Supported by recent legislative, administrative and judicial initiatives, environmental regulation in India has become very powerful. The new regulations cover noise, vehicular emissions, hazardous wastes and chemicals, hazardous microorganism and transportation of toxic chemicals. Stringent penalties were introduced in the earlier pollution control laws. The licensing regime which is supplemented by citizen suits provision together with a statutory "right to information", now enable an aggrieved citizen to directly prosecute a polluter after examining the government records and data. The technology forcing deadlines issued under the Central Motor Vehicles Rules, 1989 compels the manufacturer of vehicles to upgrade the technology to the standards. Under the EPA, rules have been notified for mass based standards for the manufacturer of vehicles for the year 1995 and 2000. In March 1992, rules were notified for environmental auditing of all the industries which may cause water or air pollution or generate solid or hazardous wastes. Workers participation in plant safety was made mandatory and stringent penalties on high level management for breach of factory safety regulations are expected to reduce industrial accidents. The vesting of enormous administrative powers in the enforcing agencies is also an encouraging step towards the improvement of the environment. A gigantic project like Ganga Action Plan is also a sign of high concern for the environment. Recently, the Ministry of Environment and Forests has adopted a "Pollution Abatement Policy" which includes the adoption of clean technology, conservation of resources, change of concentration based standards to mass based standards, incentives for pollution control, public participation, environmental auditing and Eco-mark environment friendly products. Lately, the judicial power has been reactivated through public interest litigations, not only with a view to focus attention on the problems of pollution but to check the same through the use of the existing infrastructure available in the system. In the urbanized segments of society the use and exploitation of resources of local bodies are pressed into service to tackle the urgency of the matter.

LACUNAE IN ADMINISTRATIVE POWERS OF ENFORCEMENT

Environmental legislation has now come up as a complete code through numerous legislative drills conducted by Parliament. Laws pertaining to water, air, noise, and environment in addition to several rules promulgated thereunder find a definite place in the statute book. Rules now exist to control pollution in the manufacture, use, storage, etc. of hazardous chemicals, micro-organisms, and genetically engineered organisms or cells.

In spite of all the duplication of law, there are lacunae not only in the law, but in some cases in the administrative powers of enforcement as well. This is evident from the case of "plastic carrying bags". There is an on-going debate on the health and environmental hazards caused by these bags. They not only choke drains but leach toxic chemicals into food products. The colours used in these bags contain metals such as cadmium and lead which are extremely dangerous to health. Municipal officials say plastic bags are mostly been responsible for choking drains. Ragpickers are hardly able to remove them. Some experts say, these bags cannot be recycled either. This was one of the reasons why Himachal Pradesh government wanted to make the state plastic free.

Some of the States which did impose a ban on plastic bags have discovered that the menace cannot be easily removed by merely imposing a ban on their dumping. The measure is already proving to be ineffective in the absence of any effective environment laws. In Himachal Pradesh, the first state to impose such a ban a few years ago, plastic bags continue to be littered everywhere. When the government of Himachal could not succeed in banning plastic bags, it appealed to the Union environment ministry to issue a notification under the Environment Protection Act (EPA), 1986 to do something about it, as Polythene is an important environment problem and not merely a sanitation or forest-related hazard. However, the Union environment ministry expressed its inability to do so under EPA as plastic waste is part of municipal waste, which comes under municipal acts and not under the Centre, so the Ministry of Environment feels that it should not be part of EPA.

Although more states have followed the Himachal Pradesh's example of imposing a ban under a garbage control Bill, experts believe the results will not be different. Haryana and Sikkim imposed such a ban recently. Jammu and Kashmir announced their willingness to follow the example set by Himachal Pradesh. However, the experts feel that the law needs to be more stringent to be really effective.

REMEDIES

When the administrative authorities do not exercise their power and responsibility, the judiciary has to step in and set right the issues. Various fora for settling these disputes exist under different laws including the Constitution of India. There are criminal remedies under the Criminal

Procedure Code (Cr.P.C.) and the Indian Penal Code, remedies to get a permanent or interim injunction under C.P.C., remedies of prosecution under the Environment (Protection) Act and the eponymous legislation establishing environment tribunals in the country. This is a part from the extraordinary jurisdiction to the superior court to set right the environmental issues if they impinge adversely on Fundamental Rights. The Public Interest Litigation movement was created by this jurisdiction. As far as the efficacy of the above fora is concerned, the choice of forum will depend on the remedy that is sought as well as the resources available to the litigants. If the objective is to seek an injunction against the repetition or continuance of conduct of causing damage to the environment, then a writ petition under Article 226 of the Constitution may be speedy enough. But as is well known, this envisages that the petitioner has adequate legal and intellectual support as well as financial resources to attain the objective. In the case of a minor or localized problem, it will be more helpful if judicial assistance is sought under Section 133 of Cr.P.C. If the matter is not very urgent, and is of a local nature—only district-wise importance, proceedings under Section 91 of C.P.C. may be considered as adequate. Too frequent an invocation, however, of the writ jurisdiction for minor local infringements of environmental legislation will clog the dockets of the High Court itself. Criminal prosecutions, should be resorted to only as last resort because: (1) They can be counter-productive and they can freeze the cooperation of concerns which are prosecuted—particularly if those concerns many; (2) The burden of proof is heavier in criminal prosecution wherein a number of constitutional rights are also involved; (3) As experienced in many other countries, which are quite serious about environmental problems, a cautious and guarded use of criminal prosecution yields good dividends in the long run, rather than too hasty or frequent recourse to such prosecution.

Another remedy is to set up standards, which can be used to stipulate how much of the pollutants can be safely discharged into the environment.

In principle, an economic efficient level of pollution can be defined as that level of pollution at which the marginal net private benefits of the polluting firm are just equal to the marginal external damage costs. Because of data deficiencies and the limitations of this approach, the optimum situation is not a practicable policy objective. Instead, society sets acceptable levels of ambient environmental quality, and policy instruments are directed at these standards. The analytical task is to seek out the least-cost policy package sufficient to meet acceptable ambient quality standards. Many economists favour the use of effluent taxes—per unit of pollution—but the actual pollution control policy has been based on regulatory approach often involving uniform reductions in pollution emissions across classes of industry. Despite the current vogue for suggesting an economic or fiscal mechanism for combating an environmental problem, the system of regulation by public bodies remains the prime tool for environmental protection. Regulatory modes imply that administrative regulation is far more than just the setting of rules on what can and cannot be done. It denotes a coherent system of control in which the regulating body sets a framework for activities on an on-going basis, with a view to conditioning and policing behaviour as well as laying down straight rules. The purpose of environmental regulation is to prevent, mitigate or remedy environmental damage. Such damage can ensue from the use of products (such as plastic bags) or from processes and methods of production.

Environmental standards differ also in authority upon which the standard is fixed. The environmental standard may be fixed by statute, decree or other type of government regulation. Given the technical nature of standards, however, this is seldom the case. Instead, most environmental statutes lay out only a general framework, and establish an administrative agency empowered to set the environmental standards. Again, the precise role of this environmental agency can take different forms. It can set standards or decrees that are binding (e.g. all firms of a certain branch of industry), it can regulate specific activities by requiring licenses subject to given environmental standards for that activity, or it can set customized standards by requiring individual licences subject to individual standards for each. The advantages of such a system include the ability to provide uniformity, rationality and fairness between those who are regulated. Some form of public accountability is also produced by having a public body responsible for regulation.

Approaches using combination of both statute/decree regulation and agency defined standards are also used. Both systems have advantages and disadvantages. One advantage of standards set by an administrative is then the agency has high level of technical expertise and has ability to consider local and individual circumstances in issuing licences. A disadvantage, however, is that potential inequalities and differences in licensing requirements might distort market conditions. Moreover, over time, due to lobbying and other influences, the administrative agency may become subservient to the regulated firm. In those cases, the private interests of the firm become too strongly reflected in licence conditions, and complete internalization of environmental costs does not occur.

Economic, trade and other government agencies should be mandated to develop policies that encourage sustainable development. For example, EEC countries refused to buy Indian leather, because it contained pentachlorophenol (PCP), a toxic chemical widely used as a preservative of leather and wood. The German government's new regulation virtually banned the use of this chemical in leather goods. Many alternatives to PCP exist such as Thiocyanatomethylthiobenothiazole (TCMTB), P-chloro-m-cresol (PCMC), and -phenylphenon (OPP). However, the most widely used substitute is Busan 30 which consists mainly of the chemical TCMTB. Indian producers had to replace PCP with one of the above mentioned chemicals so as to sell their products in the international market. Would they have made the changeover without the trade ban is a difficult question to answer. Another example, in the same vein, is that of the tea. It was the Germans again who banned imports of tea from India because of the high levels of pesticides. As a matter of fact, it is not only tea, but a whole gamut of agricultural products which are banned in a number of countries, because of the high level of pesticides used. Indian tea planters, under pressure from Germany, assured that all efforts would be made to reduce pesticide residue levels in Indian tea so as to meet the German standards.

CONCLUDING REMARKS

In conclusion it can be stated that there is plethora of legislation and regulating authorities—In fact, there are more than 200 pieces of legislation in India that directly or indirectly relate to the protection of environmental resources. This causes confusion and makes it very difficult for every one concerned to cope with these laws. Even though, there are plenty of laws to deal with environmental problems, there are hardly any to deal directly with solid waste problems. This can be some what explained by the fact that solid waste management is a problem of the municipal corporations and not the Central Government. But, if one sees the municipal acts they also date back to the ages when the problems caused by rapid development, over-population and growing slums did not exist. These are not only legal problems, but are sociological problems as well. Poverty is the fundamental cause which makes people over-exploit the natural resources of the country such as land, forests and water, so as to meet their basic needs for employment, shelter, fuel and fodder for their cattle. While addressing the Stockholm conference, Mrs. Indira Gandhi had correctly stated, "Poverty and need are indeed the greatest polluters". Together with poverty, there are other issues such as awareness, resources, and political will which are some of the other issues which must be kept in mind while making the laws and devising methods for their enforcement.

Section 2

Country Case Studies of African Continent

5

Introduction to Waste Management in the African Continent

Santosh Kumar[1], R. Bappoo[1] and M.B. Sasula[2]

[1] Department of Applied Mathematics,
National University of Science and Technology, P.O. Box AC 939,
Ascot Bulawayo, Zimbabwe
[2] Cleansing Section, Bulawayo City Council, P.O. Box 1946,
Bulawayo, Zimbabwe

INTRODUCTION

Africa is a vast continent with arid and semiarid areas where the landscape has been badly scarred by the exploitation of mineral resources. The period 1500-1940 was very significant in terms of agricultural and industrial development that brought about the cultural and environmental change in Africa.

The most significant environmental change in Africa was due to deforestation and the spread of agriculture. The annexation of some African territories and the imposition of European traditions and agricultural practices on indigenous people contributed a lot of confusion on the continent's environment. The environmental impacts of agriculture such as salinization and soil erosion made the continent poorer with many countries unable to sustain their environments. Most of the continent is very dry with erratic rainfall. Countries along the Equator do have good rains but they have a badly exploited ecology. Because of the Global Environmental Change, the continent is experiencing very difficult times. Poorer nations are getting flooded while some are faced with droughts and famine. The diseases associated with the above are taking their toll.

SOLID WASTE AND ENVIRONMENTAL MANAGEMENT

The waste management business in Africa is generally low profile and plays second fiddle to politics and regional boundary conflicts. In some countries like Burundi, the need to protect the environment has been overtaken by the desperate need for land to bury the dead. Countries with a better economy like South Africa, Botswana, Cote d'Ivoire, Egypt, Morocco and Libya have problems of inadequate planning and the use of inappropriate technology which have resulted in serious wastage of resources.

WASTE MANAGEMENT IN AFRICAN COUNTRIES

Because of the need to develop faster on borrowed ideas and technology, most developing African countries have ended up as dumping grounds for hazardous and toxic waste from the developed world. The African countries reflect very little concern about the possible environmental and ecological damage caused by poor waste management policies. Such crises in African cities was best summed up by Mateo Magarinas de Mello, an environmental lawyer from Uruguay in 1983, as: "anybody can throw anything, anywhere at any time and in any quantity" (International Institute for Environment and Development, 1987). Diseases that are caused by poor solid waste management are at the moment causing havoc in central and eastern Africa.

There is no preparedness in most African countries when disasters strike, the example, the El Nino effects, the frequent cholera outbreaks in Kenya, Ethiopia, Somalia, Tanzania, Uganda and Mozambique, and the Ebola epidemic in the Democratic Republic of Congo (Zaire) and Gabon.

SOLID WASTE MANAGEMENT IN SOME AFRICAN CITIES

Lusaka, Zambia

Once very rich in copper resources, the capital of Zambia has its solid waste management system in dire straits. The streets are heavily littered with domestic refuse, drains are choked and there is no form of pest control in place. The causes of poor waste management in Lusaka are due to shortage of proper refuse removal vehicles, financial constraints and lack of technology. The country's economy has become weak due to the mismanagement of the mineral resources.

Nairobi, Kenya

Solid waste management which had improved over the years has been badly affected by natural events due to the lack of preparedness and to political crises. The rich tourist resources have been compromised by the lack of environmental protection. Whilst technical staff has been trained in waste management, the political arm has made it very difficult for the relevant authorities to forge ahead with a protracted waste management system.

Pretoria, South Africa

Because of the sound economy of the country, the city of Pretoria has put in place a solid waste management system which caters well for all its residents. The Environmental Protection Plan has been put in place and solid waste management has greatly improved. Guidelines in refuse disposal are being monitored to protect the environment.

Gaborone, Botswana

The diamond rich city has started to cater for to Environmental Protection. Part of the diamond profits has been earmarked for the City's development and environmental protection. Very expensive refuse removal vehicles have been acquired but with very little technical expertise to handle or service them. This fast growing city has done very well to recycle its resources for environmental protection.

The African countries must put their heads together to improve their resources and protect the environment just as the European countries have done. Ideas or technology should be assessed before being imposed in local situations. While the Rio Summit in 1992 has made a lot of countries aware of the environment and its complexities, more practical strategies should be put in practice to safeguard the environment.

THE WAY FORWARD

The continent may progress if the Organization of African Unity (OAU), with the help of World Health Organization (WHO), envisages a Waste Management Policy for all the African countries such as developing a Waste Disposal Plan for the continent or a Regional Waste Disposal Plan and prepares a comprehensive plan for each country to deal with waste through both the public and private sectors.

6

Solid Waste Management in East and Southern Africa: The Challenges and Opportunities

Nonita T. Yap

Associate Professor,
College Faculty of Environmental Design and Rural Development
University of Guelph, Guelph, Ontario, Canada N1G 2W1

INTRODUCTION

Solid waste management, i.e., the control of the generation, storage, collection, transport, processing and disposal of solid wastes in accordance with the best principles of public health, economic, engineering, aesthetics and environmental protection, is becoming one of the most pressing and difficult challenges facing municipal governments in *Low Consumption Countries (LCCs)* [1].

The evolution of the solid waste management problem in Africa is a familiar story: "Traditionally solid waste management has evolved as mainly the removal of municipal solid wastes by hauling them out of the city boundaries and dumping them "there". This is in conformity with the "out of sight out of mind" philosophy" (National Environment Management Council, Tanzania 1996, p. 1).

While neither the evolution nor the impacts of the problem are unique to Africa the problem assumes greater urgency in the continent for several reasons: (1) the high rate of rural-to-urban migration means faster rates of increase in the amounts of wastes being generated annually[2]; (2) the changing nature of consumption goods combined with the frequently haphazard spread of manufacturing and service industries has resulted in increasing proportion of non-biodegradable and hazardous components in municipal refuse [3]; and finally (3) the burden of managing the increasing quantities, increasingly complex and hazardous wastestreams is falling on governments already unable to deliver other equally basic public services such as health, education and transport and are beleaguered by political instability.

This chapter reviews the current situation and examines trends in solid waste management practices in east and southern Africa. It draws heavily on the literature, mostly government documents, consultants' reports and publications of non-governmental organizations, as well as the author's development work and research in the region in the last 15 years, most recently in directing a course on cleaner production and consumption for mid-career professionals in the Southern Africa Development Committee (SADC) region in August-September 1997 [4]. The paper concludes with comments on the opportunities for significant improvements.

1. THE CURRENT SITUATION

Waste Generation Rates and Characteristics

There is paucity of detailed waste generation and characterization data on African cities. Where waste data are reported, the methodologies for collecting the information are rarely described. It is, therefore, very important to use the data appropriately and with caution [5].

Estimates of per capita waste generation rates in the continent range from 0.3-1 kg daily comprising 70-90% putrescible organic matter (Government of Uganda, 1994; Blight, n.d.; UNEP, n.d.). The few studies that appear to have been undertaken with rigour, reveal patterns consistent with those in other countries. For example, South African studies show that Johannesburg residents generate roughly four times (0. 58 to 0.73 kg/ capita) as much waste daily as the residents of Soweto (0.15 kg/capita) (Blight, n.d.) consistent with the positive correlation demonstrated elsewhere between waste generation rate and income levels.

Likewise, a nationwide study, initiated in 1984 and undertaken every five years, of trends in the abundance and types of marine litter along South Africa's coastline shows the increase in the non-reusable, non-biodegradable component of refuse with changing consumption styles. The 1994 survey showed that over 80% of litter is plastics, mostly food packaging, one-time use items and recreational fishing accessories (Ryan, 1994). The latest survey also suggests that the greater Cape Town area alone dumps, through the storm drains, some *four million litter items daily*, weighing 2.5 tons[6]. A 1993 waste characterization study in Botswana sponsored by Deutsche Gesellschaft für Technische Zusammenarbeit (GTZ) suggests a parallel trend. The GTZ study shows that beverage containers constitute 90% of metallic cans used and that of 204,049 tons of wastes dumped at the landfill site in Goborone, over 80% were construction materials [7].

In Botswana, the aesthetic impacts of solid waste dumping have become so visible so as to mobilize public opinion and government action. Empty cans lie on the ground and are thrown out of car windows. The road department and the city or district councils regularly pick up such cans from road sides and dump them in small landfill sites along the roads or at the official dump sites of the city or district. Vehicle wrecks abandoned by owners are picked up by wreckers who extract valuable spare parts but end up with piles of car scrap (Conservation News Botswana, 1995).

In Uganda the government attempted to intervene early in the process. In the early 1990s a study showed that the 10-15% non-biodegradable material — polyethylene, tins, bottles and metals — in municipal refuse came almost entirely from the urban areas. In response, the government banned the importation and manufacture of non-biodegradable polythene paper (*kaveera*) but these activities reportedly continue (Government of Uganda, 1994).

Most municipal by-laws in the region distinguish household waste, construction and demolition debris (CDD) as well as waste arising from industrial and medical institutions and prescribe separate collection and disposal for each stream. In practice however, one finds solid waste from all sources—food scraps, beverage cans from household, schools and hotels, animal carcasses from slaughterhouses and markets, CDD, glass and metal scrap from industries, syringes and bandages from health clinics and hospitals—all collected in many dump sites (Kasondje, 1997; Osebe, 1997). The observation made by a reporter for Nairobi's *Daily Nation*, about the open dump site in the city "... pungent smoke ... perenially spews ... used syringes, discarded blood and other body fluids from part of the refuse..." can apply to many cities in southern Africa and probably across the continent (Osebe, 1997).

Waste Handling and Collection

The United Nations estimate the collection rate across the continent to range from 20 to 80% with a median range of 40-50% (UNEP, n.d.) [8].

Generally, refuse is dumped on the ground at different points of the city. Where streets are regularly swept, the street sweepings are also gathered at the same point. These sites are then picked up by motorised vehicles, tractor-drawn trailers, human- or animal-drawn carts and taken to the dump site. Because collection is rather irregular it is quite usual to find garbage strewn all over the streets by scavenging animals or as a result of heavy rain. Heaps of uncollected waste on the streets, in alleys, by the roadsides, or in vacant lots, is a common sight. From Johannesburg to Kampala, the higher income areas are invariably better serviced than the low-income neighbourhoods.

Residents in unserviced areas bury or burn their waste or dump them in open spots. With the continually expanding and unplanned low-income settlements in many cities, increasing volumes of waste remain uncollected, blocking storm drains and causing floods during heavy rains, with the organic matter creating rich grounds for flies, mosquitoes, vermin and rodent infestation.

Kampala is perhaps not atypical. The City Council provides bins at designated collection centres. The bins, which are open communal containers, frequently overflow and invite scavenging by people and animals. Collection service is provided to less than 10% of the population and removes less than 20% of the estimated 760 tons generated daily. In the low-income areas access is frequently blocked by garbage and the piles become breeding ground for rats. This practice appears to be a radical departure from that in rural areas where it is reported that almost all rural households dispose of their garbage, 90% of which is vegetable matter, by burying them in home gardens. Very few burn their garbage (Government of Uganda, 1990, 1994).

The conditions in Dar es Salaam may be worse because of its greater population and much higher rate of urbanization (approximately 20%). The city manages to collect only about 13% of the daily refuse generated, admittedly a considerable improvement over the 3% collection rate reported in 1992. Collection however is irregular and confined to town centres and high-income neighbourhoods (UNDP, 1992; EPL, 1994).

The failure to service low-income neighbourhoods has a familiar explanation—lack of political power and low revenue potential of the poor. Poor or lack of access roads to the housing areas, frequently cited as part of the explanation, is in fact part of the problem (Kasondje, 1997; Osebe, 1997).

However, there is no question that the quality and the level of existing services to city centres and high income areas are far from adequate. Municipal refuse collection and disposal systems in most cities are plagued with insufficient funds to maintain, replace, and increase the vehicle fleet. It is estimated that at any given time, *three out of four* refuse collection vehicles are not in operation because of mechanical breakdown, lack of spare parts, lack of money to buy petrol, or are impounded by garage shops because of nonpayment for repair work. Many of the vehicles are unsuitable for the conditions of the roads. There is also inefficiency resulting from inappropriate collection times. The problem is compounded by the poor supervision of workers by the management. Waste workers are poorly paid and poorly equipped, (i.e., many do not have boots, gloves, shovels and rakes), and are therefore poorly motivated[9]. This is exacerbated by low social esteem for the job (see, for example, National Environment Management Council, 1996).

Reduction, Reuse, Recycling and Recovery

The notion of reducing individual consumption is understandably, rarely discussed in the continent, with the singular exception of South Africa, where some public debate has been generated around reducing the use of plastic shopping bags through a system of incentives and disincentives (Ryan *et al.*, 1996).

Reuse and recycling are a different story. As is typical in *Low Consumption Countries*, reuse and recycling are widely practiced in low-income households and are linked to the informal economy. Where source separation in high-income households is done it is done so by domestic servants. Where such channels exist, materials of value such as glass bottles, plastics, paper or cardboard are either reused, recycled or sold to individual middlemen or at commercial outlets where such channels exist. It is not uncommon to see waste pickers, mostly women and children, at dump sites. This frequently gives rise to tension between waste workers and waste pickers [10].

Formalizing or organizing the secondary materials collection and marketing activities enhances the benefits to the participants by reducing demands on them and perhaps improving their ability to negotiate the price. It can also increase the collection rate. In Botswana, for example, linking local can collection to the *Collect-a-Can* network has led to an almost 55% recovery rate for cans. South Africa has gone the farthest in this direction. The Khayelitsha Business Association in Cape Town established a War on Waste Centre, and helped to create 300 jobs for the local unemployed residents of the township. The Centre collects waste paper from shopping centres on a permanent basis, buys paper from 100-300 suppliers, and directs those with plastic, glass and metal scrap to appropriate buyers such as *Rhino Plastic, Consol Glass*, and *Collect-a-Can* . In the western Cape, it is reported that 22 000 unemployed people are making a living off the collection of recyclables. Some communities use the proceeds to build community facilities as has happened in Simon's town where the community used the revenues to build a community swimming pool. Post-consumer paper recovery in democratic South Africa has increased by 33%-38% in the last five years (Ntamnani, 1997).

Materials recycling and recovery, specifically of glass, paper, metals and more recently plastic is undertaken by the private sector. This is to be expected since this end of the chain is generally technology-, and therefore, capital-intensive. In Zimbabwe, a lead battery recycling plant has been in operation for 35 years. Over 75% of lead used in its annual production of about 240,000 units of standard automotive batteries and 16,000 units of industrial batteries comes from the old car batteries it purchases across the country. A waste paper collection arm was also established in Zimbabwe several years ago by the pulp and paper industry. The *National Waste Company* purchases wastepaper from across Zimbabwe. *Bata Shoes*, Malawi produces a line of colourful sandals from scrap rubber.

Surprisingly, some of the low-technology, low-cost materials recycling and recovery options, specifically the conversion of garbage to compost and/or to animal feed, appear to be the most marginalized option in southern Africa. This is rather surprising given the tremendous waste reduction that can potentially be achieved with the high "compostable' content reported for most solid waste streams. In some suburbs of Durban, Johannesburg and Pretoria there are community centres composting garden waste for use in household gardens. Backyard composting is being promoted by NGOs in Benin, Cameroon, Ghana, Kenya and Uganda (Government of Uganda, 1994; UNEP, n.d.). However, it is probably fair to say that most African cities have no significant municipal-level garbage composting or recycling scheme.

Waste Disposal

The most common practice is the use of landfills. These landfills however are not lined, and few are fenced. In Botswana, out of the 75 disposal sites reported only two are sanitary landfills (Matsoga, 1997). Of the 1200 landfill sites in South Africa only 20% are reportedly legally permitted (Ntamnani, 1997). In some cities, e.g., Harare, compaction is done but there is no application of soil cover. They are essentially open dumps. Most are located just outside the urban centres in vacant lots with close proximity to ecologically-sensitive areas. In Harare and Dar es Salaam the dump sites are close to residential areas. In Nairobi, abandoned quarries in the city have become the dumping sites for municipal refuse (Osebe, 1997). Waste dumps are frequency sited based on easy accessibility to collection vehicles rather than ecological or human health considerations.

The poor conditions and inadequacies of existing waste disposal facilities make incineration a very seductive option, particularly when offered through overseas development assistance (ODA). Indeed an incinerator-cum-energy recovery has been installed in Tanzania with foreign assistance but overall, incineration has not been a serious option for municipal solid waste management in Africa. A quick-and-dirty *technology assessment* of incineration would quickly and inequivocably show that it does not deserve serious consideration by policymakers genuinely seeking environmentally-sustainable and financially-viable solutions.

The environmental sustainability of using incinerators in Africa is at best dubious. Its financial viability, is equally questionable. The high putrescible organic matter content reported for most waste streams would make incineration a net consumer rather than a producer of energy. The capital-intensity and the high level of technical expertise required for the operation and maintenance of incinerators make this technology highly inappropriate and an unwise investment for most of Africa.

Institutional Framework

Contrary to what is frequency implied by the existing literature, there is no policy vacuum with regards to the management of solid waste in Africa. The municipal refuse byelaws of Lilongwe and Blantyre, the two largest cities in Malawi, prescribe the size and shape of the receptacles, the condition under which these receptacles are to be maintained, frequency of collection and the distances of the collection points from the residences. So do those of Dar es Salaam. Some regulations such as those of Port Elizabeth in South Africa, even specify the hours for collection (Port Elizabeth Municipality, n.d.).

It is true, however, that many of the policies are outdated. For instance, solid waste management in Harare is governed by municipal byelaws dating back to 1979. This is being remedied in some countries. National legislation and municipal by-laws are being updated. In South Africa, 36 pieces of legislation relate to waste disposal but none set any criteria for environmentally-acceptable development, operation, monitoring and closure of landfills. Such standards were established in 1994 (Republic of South Africa, 1994).

Lesotho's Environment Bill of 1997 created the Lesotho Environment Authority with the mandate to establish "standards for waste, waste classification, and analysis, and formulate and advise on standards of disposal methods and means for such waste" (Government of Lesotho, 1997). In Zambia, regulations relating to waste management were passed in 1993 under the Environmental Protection and Pollution Control Act (Act. no. 12 of 1990). It established standards for transporters of waste and operators of waste disposal sites. Also in 1993, the by-laws of Dar es Salaam City Commission, were revised allowing the delegation of municipal refuse collection and disposal to the private sector.

In Botswana, a waste management legislation, drafted with the support of German development assistance, is awaiting approval by Parliament. This legislation establishes standards for landfills design and operations and recommends the creation of a government agency to implement it. It incorporates *precautionary principle, polluter pays principle,* concept of *Duty of Care*[11], and the 4R's (Conservation News Botswana, 1995).

Agents of Change

The responsibilities for municipal refuse collection and disposal rest with the municipal councils. The service is directly managed by the Officer of Health although the collection vehicles are under the control of the municipal engineer. With few exceptions such as Dar es Salaam, Cairo and some cities of South Africa, the costs of garbage collection and disposal are borne by the municipal councils.

As described in an earlier section the private sector is involved in larger scale materials reuse, recycling, and recovery. With the democratization of South Africa, the secondary collection of materials and marketing channels are likely to gain economies of scale through regionalization. A regional network of can collecting activities has been established by *Collect-a-Can Pty Ltd*. The network spreads over southern African countries—Namibia, Botswana, Zimbabwe, Mozambique, Lesotho, Swaziland and RSA. *Collect-a-Can* is financed by the canning and steel industry of South Africa.

A voice that is only beginning to be heard on solid waste issues is that of environmental NGOs. Outside of South Africa, very few environmental non-governmental organizations (ENGOs) in the region identify urban environmental quality as a major arena for engagement. The few that do are beginning to make a difference. In Botswana, *Somarelang Tikologo*, a conservation group is vigorously pursuing public awareness and policy advocacy campaigns around waste and the 4R's. In Zimbabwe, *Environment 2000* is almost totally dedicated to urban environmental quality issues. Under its Recycling and Anti-littering Programme (RAP), it has established semi-permanent structures to serve as one-stop points where the public can deposit all recyclable wastes (RAP Centres), or direct purchasing outlets for recyclables (RAP buyback Centres). Both are operated with proceeds from the sale of the materials, either by community organizations or entrepreneurs. The RAP also places receptacles in institutions, e.g., schools, where recyclables can be deposited. The proceeds from the sale of the materials accrue to the institution [12].

It is interesting, although perhaps not surprising, to note that in most of the materials collection reuse and recycling activities, women and women's groups are either leading the initiatives or are active participants. In many countries of southern Africa, women shred, clean and reknit rags into sweaters and cardigans for resale along the roadside or in flea markets. In the village of Seleka in Botswana, it is reported that three out of every five women are involved in cutting plastic bags and weaving or crocheting them into baskets, floor carpets and other items. The entrepreneurial activities of the village women go back to the drought of the early 1980s. They reused the cloth sacks used for the 'emergency' maize meal to make bed sheets, table covers, seat covers, shopping bags, school bags, clothes and curtains (Masilo, 1996).

A very important force for change on waste management is the donor community. Perhaps more so in Africa than in other developing regions. Bilateral and multilateral development assistance agencies—Deutsche

Gesellschaft für Technische Zusammenarbeit (GTZ), Japan International Cooperation Agency (JICA), US Agency for International Development (USAID), the World Bank, the African Development Bank (AfDB)—have been funding programmes directed at improving the management of municipal solid waste in Africa for over 20 years. They have traditionally emphasized the provision of refuse collection vehicles primarily for servicing city centres (UNEP, n.d.). More agencies are getting involved e.g., the Canadian International Development Agency (CIDA), Danish Cooperation for Environment and Development (DANCED), and some are beginning to fund urban environmental initiatives of non-governmental organizations, such as *Environment 2000* (Zimbabwe) , *Somarelang Tikologo* (Botswana), and *Environmental Monitoring Group* (South Africa).

EMERGING TRENDS

There are emerging trends observed in southern Africa that may be generalized for the continent:

1) The costs of solid waste management will rise dramatically within the next five years. The first generation of radios, television, tape recorders, ovens, air conditioners, and refrigerators imported into most African countries will join the plastic and metal scrap heap early next century. The GTZ study in Botswana provides a glimpse of what the next decade could look like. The study projects that by the year 2005 and assuming a 25% by weight recovery of spare parts, an estimated 32,367 tons of scrap will be generated annually in Botswana from the 1986-1996 motor vehicles fleet. It is also projected that 2.2 tons of metal and plastic scrap will be generated annually from existing household appliances and 735 tons annually from electronic equipment.

2) The increasing penetration of the cash economy in rural areas will lead to increased consumption of processed foods. The vast distances involved and the poor conditions of the transport infrastructure will promote the use of cans, rather than bottles. The strong lobby for the increased use of aluminium could change the scrap composition and reduce the current dominance of tinplates.

3) Although the environment and human health impacts of improper handling, collection and disposal of solid wastes are most direct, acute and visible in low-income areas, the cumulative effects of inaction— poor air quality, unsafe drinking water, contamination of the food chain are starting to affect and concern high-income earners. The impacts of indiscriminate waste dumping on important economic sectors such as tourism, are also becoming obvious to governments. These will eventually drive efforts for genuine improvement.

4) The privatization of solid waste collection and disposal services is seen as an important part of, if not the solution to the solid waste management problem . The donors are pushing for it and indigenous private sector capacity is growing. In Dar es Salaam, for example, *EPL,* a private sector enterprise has, since 1993, been involved in solid waste collection and cesspit drainage services in the high-income areas of Oyster Bay and Msasani. It has been collecting the garbage from more than 10 embassies and 200 residential and commercial premises. *EPL* was recently awarded a contract by the Dar es Salaam City Commission (formerly Dar es Salaam City Council) to provide environmental services in the low and high density periurban areas of the city, comprising 28% of the total city population with an estimated refuse generation of 400 tons daily (EPL, n.d).

5) Community education is increasingly recognized as indispensable even by private sector enterprises. Dar es Salaam-based *EPL* is planning to involve ward, village and "cell' leaders in its solid waste management plan in the six wards it has been contracted to serve. It plans to use the mass media—radio, television, newspaper and brochures—to inform residents of collection routes and schedules of pick-up vehicles as well as to educate them on their responsibilities with regards to payment of collection charges (EPL, n.d.).

6) The liberalization and regionalization of African economies will inevitably lead to *policy permeability* of national borders. The national governments will be less and less able to institute independent policies. Environmental protection and resource conservation strategies will require collaboration and cooperation across the region if they are to be effective.

CONCLUSIONS

As daunting as the solid waste management challenge appears to be in southern Africa, the problem is not intractable. There is no doubt that tremendous financial resources will be needed to achieve a significant breakthrough in the problem. Because of the generally low level of overseas direct investment (ODI) in Africa donor assistance is indispensable. However, it would be a big mistake to assume that all that is required is a massive dose of funds.

Effective and sustainable solutions will require strategic analysis of the problem. Three questions need to be asked: (1) Where can the biggest gains be made? (2) How can these gains be achieved and sustained under the conditions that are prevalent in Africa? (3) Who can best deliver the results at the least cost?

Where can the Biggest Gains be Made?

A systematic analysis will reveal that the biggest gains at the least cost, i.e., biggest reduction of waste quantities and risk per dollar invested, can only be made in the front end of the chain. The high content of putrescible organic matter reported for most African urban waste would strongly suggest that wastes going to the dumpsites could be reduced by at least 50%, by recycling the organic matter as animal feed and/or by composting it for backyard gardens, or for 'green zones'. The non-biodegradable component could be dramatically reduced through materials reduction, reuse, recycling and recovery.

How can these Gains be Achieved under Local Conditions?

The poor constitute the urban majority. They undoubtedly create part of the solid waste problem and therefore need to be part of the solution. Environmental solutions must benefit, and be seen to benefit the poor. Environmental problems and solutions should be linked to their livelihood and quality-of-life concerns. The link between indiscriminate waste dumping and health should be demonstrated. To bypass the barriers of language and illiteracy education programmes should use popular education techniques—theatre, folk songs, story telling and so forth.

Environmental issues could be articulated within the traditional value system and if appropriate, organizational models should be anchored on traditional authority structures.

Matsoga's description of traditional beliefs, ceremonies, and sanctions in Botswana with regards to maintaining cleanliness in the residential areas while cursory, clearly points to the need for revisiting traditions, e.g., conservation ethics, cleanliness norms, so that they might serve as building blocks in redefining society-environment relations in the 'modern' context[13].

Assaad's study of the transformation of Cairo's refuse collection system is instructive with regards to the restructuring of traditional institutions to meet "modern" needs. He describes how confrontation was averted between Cairo's municipal authorities who were determined to "modernize" the city's refuse collection system and the *wahiya-zabbaleen* communities who had traditionally provided the service through an internal albeit informal set of rules, rights and sanctioning mechanisms. The case demonstrates what has been shown in other places such as India, that traditional structures can be reshaped to respond to "modern" demands[14].

Who can Deliver the Results at the Least Cost?

The push for privatization needs to less ideologically-driven and more empirically-based. Capital-intensive, "modern", mechanized and "efficient" systems may be needed to meet some needs; other systems—labor-intensive, community-or family-based enterprises may be more effective and "efficient" for other parts of the city. The managed evolution of the *zabbaleen-wahiya* system shows that flexible and diverse institutional arrangements are needed and can be created, to deal cost-effectively with the diversity of needs in an urbanizing environment (Assaad, 1996). Local enterpreneurs, women's groups, educators, artistes, primary health care givers, labor unions—they all need to be involved.

Programmes need to be appropriately targeted. They need to be inclusive, integrated, and, adapted to indigenous value systems, structures, capacities and resources. Obviously, strategic alliances are needed.

END NOTES

[1]   The term *Low Consumption Countries* refers to countries characterized by low levels of consumption of basic needs. Basic needs are defined as food, shelter, education, health, choice, and political stability.

[2]   The urban population growth rate has been estimated to range from 7% in Malawi to 21.7% in Tanzania.

[3]   The term "municipal solid waste" is used here as reflected in the literature reviewed. The term includes refuse from households, industrial and commercial establishments, hospitals, markets, and yard and street sweepings.

[4]   The author was director of the workshop "Targeting Sustainable Development through Cleaner Production and Consumption" held in Harare, Zimbabwe, August 18-September 2, 1997.

[5]   A consulting firm contracted to undertake one such study attached the following disclaimer in the Report: "This report if confidential to the client and...Ltd. accepts no responsibility of whatsoever nature to third parties to whom this report, or any part thereof, is made known. Any such party relies upon the report *at their own risk*" (italics added).

[6]   This contradicts the claim that 95% of waste generated in South Africa is landfilled, 2.5% recycled and the rest illegally dumped (Republic of South Africa, 1997).

[7]   This should not, however, be taken as representative of what is actually generated since a certain degree of reuse and recycling of materials does take place particularly in low-income households and any putrescible organic matter would likely have decomposed or have been scavenged at the roadside "collection points.

[8]   This appears too high an estimate from the experience of this author.

[9]   In Dar es Salaam, for instance, waste workers receive less than 30% of the minimum wage (National Environment Management Council, 1996).

[10] Interestingly enough, the waste pickers appear to enjoy very little support from waste management professionals. At the ECEP workshop on Cleaner Production and Consumption in Harare attended by 40 waste management professional in the SADC there was a heated debate about waste pickers. Most view the existence of this informal sector as "immoral".

[11] *Duty of Care* is probably equivalent to the principle of *Due Diligence* in Canada.

[12] From Charlene Hewat, Executive Director of Environment 2000, in a lecture at the ECEP Workshop on Cleaner Production and Consumption, Harare, August 18-September 2, 1997.

[13] Matsoga (n.d.) writes of the use of *Dithotobolo*, special areas designed as dump sites, of the inspections undertaken by Kgose Bathoen II of Bangwatketse to check on the cleanliness of the community, the use of regiments to clean sections of villages and the penalty imposed for violation (e.g., strokes on the back or fines in the form of livestock or civil service). The erosion of community cleanliness norms apparently came with the differential provision of sanitary services by the British colonial government along colour lines and income levels. The perpetuation of the practice by the post-colonial government contributed to the further decline. This does not preclude governments and community-based organizations from revisiting, reviving and reinstitutionalizing appropriate aspects, or building new approaches on the old principles and concepts.

[14] A study of the mandal system in India reached a similar conclusion (Rahman and Prasad, 1995).

REFERENCES

Assaad, R. 1996. Formalizing the informal? The transformation of Cairo's refuse collection system. *Journal of Planning Education and Research,* 16: 115-126.

Blight, G.E. n.d. Sanitary Landfill Requirements. A paper presented at the Seminar on Sanitary Landfill Design, Management and Operation, Manila.

Conservation News Botswana. 1995. Vol. 2 (2). June.

Dar es Salaam City Council By-Laws. 1982-1993.

Government of Botswana. 1996. *Study on the Recycling of Metal Wastes*: Report no. NCS/GTZ 6/96. Gaborone, March.

Government of Lesotho. 1997. National Environmental Policy for Lesotho.

Government of Uganda. 1994. Ministry of Natural Resources. State of the Environment Report for Uganda.

Government of Uganda. 1990. Ministry of Water & Mineral Development and Ministry of Local Government. Deutsche Gesellschaft für Technische Zusammenarbeit (GTZ). *Solid Waste Disposal - Tororo*. Final Report prepared by Environmental Resources Limited.

Local Government (Urban Areas) Act. City of Blantyre Refuse By-Laws. Government of Malawi.

Kasondje. 1997. Solid Waste Management. A paper prepared for the ECEP Roundtable on "Targeting Sustainable Development through Cleaner Production and Consumption", Harare. September 1-2.

Masilo, R. 1996. Women and Plastic Shoppers in Seleka Village. In: *Somarelang Tikologo Newsletter,* Gaborone, March, p. 12.

Matsoga, VTC. n.d. Public Awareness Campaigns in Waste Management in Botswana.

Matsoga, VTC. 1997. Waste management in Botswana—evolutionary revolution. In: *Conservation News Botswana*. Vol. 4 (1). June.

National Environment Management Council. 1996. *Solid Waste Management in Some Urban Areas in Tanzania*. December.

Ntamnani, N. 1997. The poor make a living from waste. In: *Leading Edge: Exploring Development Growth and Empowerment*. Issue 8, July.

Osebe, A.D.M. 1997. The State and Challenges of Alternative Management Options for Municipal Refuse in the City of Nairobi, Kenya. A paper presented at the ECEP Roundtable on Cleaner Production and Consumption, Harare, September 1-2.

Port Elizabeth Municipality. n.d. Refuse Removal Regulations. Port Elizabeth Municipality. Government of South Africa.

Rahman, F.U. and M. Prasad. 1995. India's Mandal system: Its potential role in responding to environmental problems in Kanpur. In: N.T. Yap and S.K. Awasthi (eds.), *Waste Management for Sustainable Development in India. Policy, Planning and Administrative*

Dimensions with Case Studies from Kanpur. Tata McGraw-Hill Publishing Company Limited, New Delhi. pp. 181-200.

Republic of South Africa. Department of Water Affairs and Forestry. 1994. Waste Management Series. *Minimum Requirements for Waste Disposal by Landfill.*

Republic of South Africa. Department of Environmental Affairs and Tourism and Department of Water Affairs and Forestry. 1997. *Discussion Document towards a White Paper on Integrated Pollution Control and Waste Management,* May.

Ryan, P. 1994. Plastic litter in marine systems: impacts, sources and solutions. In: *Plastics Southern Africa,* November.

Ryan, P.G., D. Swanepoel, N. Rice and G.R. Preston. 1996. The 'free' shopping-bag debate: costs and attitudes. In: *South African Journal of Science,* 92: 163-165.

The 1994 Environmental Protection Ltd. (Mazingira). 1994. *Solid Waste Management in the City of Dar es Salaam. Plan of Operation 1996/1997.*

United Nations Development Programme and United Nations Centre for Human Settlements. 1992. *Managing the Sustainable Growth and Development of Dar es Salaam. Privatisation Proposal.* Sustainable Cities Programme Report on Solid Waste Management, October.

United Nations Environment Programme. International Environment Technology Centre. n.d. *International Source Book on Environmentally Sound Technologies for Municipal Solid Waste Management.* Technical Publication Series [6].

7

Solid Waste Management: A Developing Country's Perspective

S. Kumar[1], R. Bappoo[1] and M.B. Sasula[2]
[1] Department of Applied Mathematics,
National University of Science and Technology, P.O. Box AC 939,
Ascot Bulawayo, Zimbabwe
[2] Cleansing Section, Bulawayo City Council, P.O. Box 1946,
Bulawayo, Zimbabwe

INTRODUCTION

Very few countries in the developing world, especially in Africa, have really addressed the issues related to the environment. Instead, most of them have been concentrating on border disputes or wars, internal conflicts, ethnic violence, political power which have rendered their economies weak, and thus have very little funds to set aside for environmental issues. There has been, therefore very little or no interaction on solid waste management and its complexities.

The purpose of this chapter is to bring about:
1) An understanding that solid waste management is of vital importance in urbanization;
2) An awareness of the need for integration of an effective solid waste management policy in cities and towns of the developing world,
3) An understanding of solid waste management and environmental assessment policy,
4) The need to review existing methods of solid waste management and appraise them, and
5) The need for national interaction in solid waste management.

URBANIZATION

As cities grow, they experience great pressure on land, housing, water, energy, food, jobs and sanitation. This is also witnessed in Zimbabwe. The push and pull factors for the growth of cities in Zimbabwe due to industrialization and urbanization are given in Table 1.

Table 1. Push and Pull Factors

Push into City	Pull into City
Unemployment	Job Opportunities
Limited Health Care	Good Health System
Limited Educational Facilities	Good Education System
Family Pressure	Better Transport
Dependence	Industrial Development
Food Shortages & Drought	Better Living Conditions
War and Politics	Better Housing Facilities
Social Problems	Economic Depression
Static Status	Social Advancement
Over-crowding	Physical Attraction

Unemployment is one of the biggest problems faced in Zimbabwe. In a country where about 80% of its population live in the rural areas and where the rate of unemployment stands at about 40%, there is likely to be a continuous flow of people moving into the cities and towns looking for jobs. This has resulted into the problems of squatters in and around the big cities, namely Harare and Bulawayo.

The unavailability of productive lands to resettle people who have been displaced from their rural homes during the liberation struggle, the recent drought of 1991-92 associated with food shortages, overcrowding in the communal lands, lack of basic health care and education, the transport problem and the status quo in the state of development in the rural have worsened the situation. People are not prepared to go and live in those areas.

This has resulted into aggravating the following problems faced by the cities [Bappoo and Kumar (1995)]:
a) a marked increase in the population growth rate (6% annually),
b) services getting poorer due to inadequate infrastructure,
c) solid waste management poor budgets gets poorer,
d) lack of funds resulting in lack of expertise in Solid Waste Management,
e) lack of financial resources have often resulted in purchasing the wrong type of vehicles or equipment.

Thus, solid waste management can no longer be considered in isolation in a developing country, as the next century will see more people living in the cities and towns than in the rural areas.

The problems pointed out in this section are likely to multiply if these are not addressed to now. In fact, as cities are growing, the Solid Waste Management has to grow in respect of not only the resources but in technology as well. It is a protean system [Kumar (1995), Bappoo and Kumar (1995)]. All associated problems can be addressed by short and long term planning.

SOLID WASTE MANAGEMENT

Solid waste management in the urban areas of developing countries is under constant public and political scrutiny, especially when the protection of the environment is now a major issue of international concern (Murerwa, 1992). It is now very vital that waste management projects which are major cost centres must be managed to high standards.

Politics plays a major part in planning, especially in the developing world, and thus we cannot escape its grip. Politics controls all issues: social, health, technical, economics, financial, environmental, institutions, etc. With proper training in the technical and management fields, politics can be used to strike a balance between urban and rural developments. That is, to provide the same facilities such as water, sanitation and basic education to the rural majority which are currently being accorded only to the urban dwellers.

OBJECTIVES

Every country, or rather a developing country, should have objectives by which solid waste management is operated. The following are a few suggestions that may be applicable in an urban area.

1) Operational Objectives

Type of waste in the waste management project must be known for developing a sound disposal policy. For example,
a) Normal domestic waste—coming from private homes,
b) Bulky waste—broken furniture and appliances,
c) Foliage or garden waste—often seasonal in some areas,
d) Street waste—arising from street sweeping. Where collection of domestic waste is poor, street waste will include a large portion of domestic waste. Street waste will include spilled loads an dead animals,
e) Market waste—generated in large quantities,
f) Drain waste—from open drains is wetter and is more offensive than street waste,
g) Commercial waste—may include large amounts of solid waste, for example, spoiled food and packaging,

h) Office waste—likely to contain large quantities of recycled paper,
i) Food waste—hotels and restaurants produce large quantities of food waste which can be fed to animals and
j) Institutional waste—may include waste from:
    i)   Hospitals which generate both domestic (from kitchens), and office types of waste, and more hazardous pathological and surgical waste from infected dressings, syringes, etc. The pathological and surgical waste must be disposed with great care.
    ii) Confidential documents which need special disposal to ensure that unauthorized people cannot see them. Drugs, pornography and condemned foodstuffs also require special controls.
    iii) Barrack, schools and churches also produce waste which is mostly a mixture of office and domestic waste.
k) Industrial waste, if not managed well, may pose a variety of problems and usage such as:
    i)   Mining and mineral waste dumping on the tip may cause instability and water pollution.
    ii) Manufacturing waste can provide useful sources of stock feed for recycling industries.
    iii) Construction waste can be used for building temporary tip-site roads, and for cover material.
    iv) Chemical waste can pollute water sources if not properly disposed. They may also be toxic (through ingestion, inhalation or skin contact) and react together to start fires or produce dangerous products.
    v)   Agricultural waste needs careful management to minimize the breeding of insect vectors and rodents so as to prevent pollution of water sources.

2) Constraints

a) Refuse storage: The refuse likely to be produced at each source should be considered when providing storage facilities.
b) Service: The level of service (frequency of collection) must be consistent with the amount of refuse produced. The mode of refuse collection— at kerbside, house to house, or communal collection—must be determined.
c) Considerations are also needed for the policy towards squatters and/ or refugees. Very few urban authorities bother about areas occupied by them. The only time such communities come into light is when their votes are needed at election time.
d) Transport of refuse: Suitable vehicles should be used for refuse collection. Precautions should be taken in the case of vehicle breakdown.

e)   Refuse disposal: This requires the study of disposal methods which are acceptable and to what degree can pollution be tolerated.
f)   Resource recovery and conservation: This requires decision concerning recycling schemes.
g)   Financial objectives: Incomes are generated from the beneficiaries. The customers must be charged as per the amount and nature of refuse produced.

3) Employment Policy

Conditions of work and incentives will determine the quality of staff in solid waste management. The welfare of the workforce is also an important factor in attracting quality staff.

4) Finance

In Africa, solid waste removal is often the most expensive urban service. Generation of the necessary revenue is often a problem because most cities and towns do not have the necessary technology for developing viable recycling programmes which could boost their solid waste management policy. The solid waste management is directly funded by the people through rates. Hence, for some developing cities, the interest and replacement on loans used to purchase vehicles and plants that are no longer operational is always an extra burden.

Some income to finance the solid waste management is obtained from the sale of materials and compost. In Bulawayo, compost is made at compost plants from recyclable materials like grass, offal remains, sawdust and leaves. But the amount received from the sale of compost is generally far less than the cost of production.

In some Sub-Saharan urban areas like the city of Bulawayo in Zimbabwe, salvaging materials from the municipal landfill site is done through the system of tenders. A good number of tenders is won by the informal sector groups (jobless people). One such group is made up of former squatters or scavengers who are now making a better living from salvaging plastics. Usually the tender is given to the highest bidder, but not much funds are derived from this system as very few people and organizations are interested in this venture.

5) Planning

Some solid waste management institutions do not have specific plans because they do not have the money to implement planning decisions. Such institutions actually spend more money because during emergency cases they are forced into making quick and often decisions from what is available. It is quite clear that if an organization is able to set aside funds each year towards replacement of vehicles when the need arises, that organization will not be forced to operate old vehicles long after their economic lives are over. The economic life of a vehicle or plant is said to have expired when it is cheaper to buy a new one than to operate an old one because the costs of operation and repair are too high. In this part of Africa a local authority acquires vehicles only when an emergency develops. This situation leads limited choice of the type of vehicle and so the authority ends up in getting unsuitable refuse removal vehicles of different makes for which spare parts and expertise are scarce. Such situations are common in Africa and hence operational cost of solid waste goes higher.

In many developing countries, it is common that equipment is chosen for the wrong reasons. Unsuitable equipment may be selected because
a) The solid waste management team concerned does not have sufficient technical understanding to make the right choice.
b) The managers with the technical knowledge do not present their arguments to their superiors forcefully or convincing enough.
c) Managers with the technical competency may not have been consulted.

Reasons (a) and (b) can be overcome by proper training but cannot escape the grip of politics. There is a great need to collect data on the performance of different systems so that decisions can be based on facts rather than on impressions. The United Nations Centre for Human Settlements (HABITAT) computer program on the selection of appropriate vehicles for refuse collection is very important in solid waste management.

6) Wages

Solid waste management is a labor intensive undertaking which absorbs a large amount of revenue. It is very important that every worker is motivated to work efficiently and so every effort must be made to encourage high productivity.

In some cultures, solid waste management may not be highly rated as a lucrative profession. This situation can be improved by offering attractive salaries and providing good working conditions. For young engineers and managers to be encouraged into careers in waste management, the salaries should be competitive with salaries that are offered in other related disciplines. In many first world countries where even the equipment is better, salaries are extremely high in this profession so to attract better personnel.

Good personnel in solid waste management are to be valued highly and paid accordingly if the investment in their training and their experience is not to be lost to other employers.

Vehicle and plant mechanics are vital to the efficient operation of collection and disposal undertakings. Most of the machinery is complex and therefore, mechanics who work on such expensive equipment should be highly trained.

7) Control of Expenditure

The high costs of labor and equipment need careful monitoring. There is great need for expertise in financial control in the management team. Astute supervision is also needed to ensure that each person gets paid for a good day's work (Coad, 1992).

LOCAL AUTHORITY

Domestic refuse collection and disposal is usually the responsibility of the town or municipality. Local authorities are obliged to provide domestic refuse collection services.

Industrial waste and other refuse can be handled by the private sector as it is the duty of the refuse generators to collect and dispose of their waste. However, in the interest of public safety and the environment, it is the responsibility of the local authority to ensure that all refuse is properly disposed of.

1) Municipal Waste

Street waste and domestic waste are linked in the sense that domestic waste that has not been collected becomes street waste. It is more expensive to sweep up waste from the street than to collect the same quantity from domestic receptacles. A better and efficient domestic waste collection service results in less streets waste. In order to strike an optimum balance, the two functions should therefore be combined. There are some advantages in combining drain cleaning with street sweeping because street refuse may be blown or swept into drains or drain waste may be left on the road verges.

2) Street Litter Bins

Careful planning is always required in the provision of adequate and durable street litter bins for use by the public: No member of the public should have to hunt for litter bins and these bins should be distinct. Very light street litter bins can be easily moved about by members of the public. Some bins are even used as seats, making it very difficult for users to deposit litter in them. This has been observed at district shopping centres and at bus terminals.

There is a need for a vigorous health education campaign to educate the public about litter. Education campaigns that include schools have always been very successful.

The collection of street litter is becoming very expensive because of its unpredictable behaviour. A sound collection system should be enforced (hit squads) on standby in case of spilled loads or vandalized storage facilities.

3) Disposal of Waste

Solid waste disposal is often one of the most expensive services that any local authority offers and the generation of necessary revenue becomes a big problem. Generally, refuse collection and tipping fees in developing countries are heavily subsidised to encourage people to use the facilities. Where the charges are too high for the communities, there is always resistance to use the facilities and thus refuse is dumped in public places where it must then be collected at a greater expense by the local authority.

RECOMMENDATIONS

1) General

It is imperative that consultants be involved to appraise solid waste management systems. Research into the solid waste management includes:
a) Collect data on refuse generation, rates, composition and density;
b) Examine existing socio-economic and cultural baseline projects;
c) Study the existing collection system and modes of transportation;
d) Examine existing disposal facilities with respect to location and capacity of landfill sites, environmental issues, fires and scavengers;
e) Consider land reclamation through engineered landfill practices;
f) Examine landfill sites close to drinking water wells or future public water supply such as aquifers, and the development of recreational surface waters;
g) Resource recovery programmes—refuse recycle with a view of refuse reduction and employment generation;
h) Look into existing equipment and their maintenance;
i) Provide basic management recommendations regarding the institutional and financial arrangements in refuse collection and disposal; and
j) Privatize refuse collection, disposal and resource recovery programmes.

2) Some Specific Recommendations Proposed to the City of Bulawayo

a) Multiple Depots and Disposal Facilities
The location of depot(s), forms the nucleus of this system. These depots are where all the vehicles are housed and operate from. In Bulawayo there is only one such depot which is located about 3 km from the City Centre. If more than one depot is created, strategically located in the western and eastern parts of the city, for example, this would reduce the distance the vehicles have to travel to their collection sites.

Having multiple depots and multiple disposal grounds can ease the problems faced in the collection of garbage and can also be one of the lasting solutions when the city undergoes changes. There will be "savings" in distance and time for each vehicle, and this could be used to increase output (task). Hence, less number of vehicles will be needed to carry out the same number of tasks.

b) Vehicle Routing and Scheduling
A vehicle's route is a sequence of pickup points, which the vehicle must traverse, starting and ending at a depot. A vehicle's schedule is the sequence of pickup points associated with arrival and departure timings. The vehicle must traverse the points in the designated order at the specified times. These routing and scheduling plans can be more easily and efficiently designed and monitored by multiple depots since each mini depot would administer only the areas allocated to it and this would reduce the pressure on the main depot to see that every thing is carried out as per plan. Task allocation can also be easily altered in the case of under utilization of manpower and vehicles, breakdown of vehicles or in case of protean networks (expansion of the collection area) to suit the required state.

c) Minimization of Vehicles
Optimum utilization of vehicles is very important so as to cut down on operational costs in the deploying of refuse vehicles to collection sites. The general concept is as follows: suppose a fleet of vehicles has been allocated to a mini depot which has n tasks $t_1, t_2, ..., t_n$" to complete per day. Each task t_i must have a fixed time of completion T_i, a duration D_i, a start collection site S_i, and an end collection site E_i. For example, if $T_i = 10{:}30$ and $D_i = $ one hour, then task t_i must begin at 09:30. Assume that the time to deadhead between any pair of points (sites) P and Q is given by $DH(P, Q)$. Then task t_i can be followed by task t_j in a vehicle schedule if $T_j - D_j > T_i + DH(E_i S_j)$. Using this approach in vehicles scheduling, the number of vehicles could be minimized while at the same time servicing each of the required tasks.

d) Breakdown of Vehicles and Maintenance
The vehicles off the roads outnumbered operational vehicles. One of the reasons for this is that the life of most of theses vehicles had outlived their

lifespan. The lifespan of a refuse vehicle is normally five to seven years. It is, therefore, important to assimilate information about the dates when the servicing or overhauling of these vehicles were done or is due so as to avoid future breakdowns.

Using all these data, one can possibly have sufficient information to design a life-table for each vehicle used in garbage collection. Hence, the City Council would be well informed in advance about any situation that could arise and plans could be made accordingly.

Another cause of concern is that the City of Bulawayo has a variety of makes of vehicles used in the collection of garbage. These poses a problem with maintenance since not all mechanics can be skilled to work on all makes of vehicles. It is also very costly to train mechanics for such diverse purposes. Moreover, it becomes extremely difficult and costly to keep a good stock of spare parts for all these different vehicles. The City Council should consider vehicle standardization, which would reduce maintenance cost. If this is not radically possible, then efforts should be made to standardize components of the fleet, for maximum parts interchange, without standardizing the entire fleet. The City Council should also consider the types of vehicles most suited for the type of refuse produced in the city. The refuse produced in the high-density areas is usually heavy and as such the compactors presently used are not recommended in these areas. The compactors are best suited in the low-density areas where the refuse is light.

e) Modernization of the Collection System
The City Council must consider gradual mechanization of garbage collection system. This could be one way to cut down on operational cost. The City of Bulawayo with a population of about a million cannot afford the luxury of continuously spending more than a million dollars every month on the cleaning of the city (Bappoo and Kumar, 1995). More than one-third of the annual budget of the Health Services Department which goes to the Cleansing Section, could be slowly cut down and channelled to primary health care facilities if alternative solutions such as mechanization are adopted.

The mechanization process might be costly at the outset. But it has long-term benefits. The City Council might like to acquire one mechanical hoist-compactor for domestic refuse collection and could be experimented in the low density suburbs (due to light refuse in these areas) and where the residents could afford to purchase another type of bin for the new collection system. Its cost-benefits should be then analyzed before taking further measures.

f) Computerization of the Cleansing Section
The Cleansing Section of the City Health Services Department bears a huge responsibility.

It plans and supervises the work being done by 628 sanitation men. It is responsible for the cleansing operations and maintenance of the sewage system of the whole city.

It absorbs a budget of more than 12 million Zimbabwe dollars annually. It therefore needs the support of the City Council to prove its efficiency and viability. It is a Section where continued research should be carried out to improve efficiency in performing the normal functions at minimal cost. Since data are very important for research in this field and need to be updated regularly, computerization of the system is most essential. The time-table and task allocations can be computerized. Routing and scheduling problems can also be easily solved by using existing software, modified, if necessary, to suit specific circumstances. Changes arising within the collection system can speedily be solved and modified to suit the new situation. Computerization will facilitate the Cleansing Officers who presently have to prepare all the work manually, and they would then be free to supervise and ensure that all functions are carried out as per instructions.

CONCLUSION

Every developing country should develop its own solid waste management policy that best suits its economy and environment. Countries that have tried to duplicate the technology designed for the western world in the waste management system have not been very successful, especially in the choice of vehicles. The types of refuse, method of refuse removal, the routing of vehicles, location of disposal facilities and recycle recovery programmes need attention in order to develop a sound solid waste management policy.

The appraisal should be broadly assessed with a view of revamping the solid waste management system. Pressure on land is being seen or felt in developing countries and every available land is being set aside for housing and industrial projects with very little attention to disposal sites. Refuse minimization and other forms of disposal other than landfilling must be explored. Chemical and other toxic wastes must be incinerated (if acceptable to the environment) and must not be allowed to find their way into domestic and litter bins.

REFERENCES

Bappoo, R. and S. Kumar. 1995. Vehicle routing for municipal waste collection: A case study. A report submitted to the Bulawayo City Council. Paper also presented at the Fifth Symposium on Science and Technology, *Towards Capacity Building in Science and Technology*, Research Council of Zimbabwe, Harare, pp. 37, Abstract 3.24. To appear in Zimbabwe Journal of Science and Technology, Vol. 1, No. 1, 1999.

Coad, A. 1992. WHO STC. *Solid waste management. A paper presented at Loughborough University, UK.*

Kumar, S. 1995. Optimization of Protean Systems : A Review, APORS'94 Fukuoka, Japan, (Eds.) M. Fushimi and K. Tone, World Scientific Publishers, pp. 139-146.

Murerwa, Herbert. 1992. Towards national action sustainable development—The report on the national response conference to the Rio de Janeiro Earth Summit. *A report submitted to the Government of Zimbabwe*, November 2–4, pp. 1-96.

8

Deployment of Vehicles from a Central Depot to Collection Sites when Each Vehicle Makes at the Most One Trip

R. Bappoo and S. Kumar

Department of Applied Mathematics,
National University of Science and Technology,
P.O. Box AC 939, Ascot Bulawayo, Zimbabwe

INTRODUCTION

This chapter deals with the optimum allocation of tasks to a fleet of vehicles of varying capacities dispatched from a central depot to a large number of collection sites. Applications of the truck dispatching problem is one aspect of the transport problem where extensive research has been carried out (Dantzig and Ramser, 1959; Clark and Wright, 1964; Beltrami and Bodin, 1974; Bodin and Golden, 1981; Golden *et al.*, 1983; Bramel and Simchi-Levi, 1995). An efficient and well organized transport system in any organization reduces the operational costs involved. Some applications of the truck dispatching problem from a central depot include the delivery vehicles, the refuse collection vehicles, goods vehicles, etc.

The truck-dispatching problem was formulated and solved as a mathematical problem by Dantzig and Ramser (1959). The method used tended to lay more emphasis on optimizing the capacity of the trucks than on minimizing the distance. Clark and Wright (1964) studied the problem of scheduling vehicles from a central depot to a number of distribution points using a heuristic method and managed to produce a better result to the problem. Later, Golden *et al.* Bodin (1983) came up with a similar method, known as the augment-merge method.

Here, we shall look into the problem of dispatching trucks of varying capacities to collection sites (residential/commercial/industrial) for garbage removal and disposal, assuming that the shortest route between any two desired points is known. The study on garbage collection where more than one depot and multiple disposal facilities exist was proposed by Bodin and Golden (1981). However, the proposed study has been motivated for considering the problem of garbage collection in the City of Bulawayo, Zimbabwe, where only one depot and one disposal facility exist (Bappoo and Kumar, 1996). In this paper, the augment-merge method proposed by Golden *et al.* (1983) and the principle by Clark and Wright (1964) are used to find an optimum or near-optimum solution to the truck dispatching problem for refuse collection.

It is assumed that the feasible shortest routes between any pair of sites in the system are known. The goal is to allocate a minimum number of vehicles in such a way that all the sites are serviced while the total distance covered is also kept to a minimum. It is also assumed that each truck makes only one trip per day. In the appendix attached the solution obtained by the method has been evaluated.

The text is divided into four sections:
1) The formulation of a garbage collection problem in a real life situation and explained here are the ideas of "saving distances" and "linking sites";
2) An algorithm for solving the problem of allocating vehicles to sites;
3) A numerical example;
4) Problems for future investigations, which deal with the protean case (Kalyan and Kumar, 1987; Kumar, 1995; Kumar and Bappoo, 1995).

1. FORMULATION OF THE PROBLEM

a) Notation and Assumption

The following notation and assumptions have been used to develop the model of the problem.

Consider a city with n residential suburbs $S_i (i = 1, 2, ..., n)$, where the number of refuse bins is assumed to be equal to the number of residential stands in each suburb. The distances joining the end-points of the suburbs are calculated using the shortest practical routes and are also assumed to be symmetric [i.e. $d(S_i S_j)=d(S_j S_i)$ $\forall i \neq j; i, j = 1, 2, ..., n$]. Assume on day i the number of sites serviced is m_i so that in a week all sites are provided the service at a required level of frequency.

$d(S_p S_q)$ = shortest distance in linking the site S_p to the site S_q.
$ds(S_p S_q)$ = distance saved in linking the site S_p to the site S_q.
S_x = the node representing the central depot.
S_y = the node representing the dump.
trip = attribution of a vehicle to a collection site/sites for garbage removal and disposal.

It is further assumed that sites $S_i(i = 1, 2, ..., m)$ can be linked to one another using the shortest route. Each vehicle is dispatched from the Central Depot to a collection site(s) for refuse removal and thereafter goes to dispose the refuse at the dump using the shortest route. Furthermore, as in a real life situation, the sites are found on different shortest route "runs". For example, all vehicles staring from the Depot (S_x) and reporting to the various sites S_i may have different shortest paths. Similarly, after collection, they all have to go to the Dump (S_y), and shortest routes from their present locations to the Dump may be different. This is illustrated in Fig. 1 [e.g., S_x to S_i is the shortest route from Depot to site S_i and, S_i to S_y is the shortest route from S_i to Dump $S_y(i = 1, 2, ... m_i)$]. Moreover, the sites are serviced, starting from the furthest site, from the Dump.

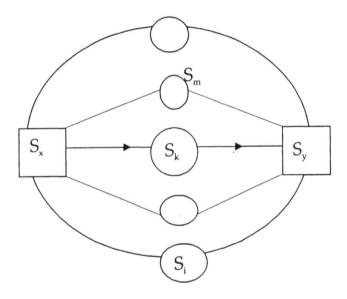

Fig. 1. Showing the Depot, collection sites and Dump

b) Idea of Linking Sites and Saving Distances

It may be recalled that the aim is to cut down on operational costs. This is done by
i)   minimizing the number of vehicles used in garbage collection, and
ii) reducing the distance covered by each of these vehicles.
    Hence, we compute the possible distances saved by linking more than one site serviced in one trip. This idea can be illustrated by an example. Consider that two sites S_m and S_k are serviced by the same truck subject to the constraint that the capacity of the truck is sufficient to serve both sites.

Then the total distance saved could be calculated as:
$$d_s (S_m S_k), = d(S_m S_y) + d(S_x S_k) - d(S_m S_k).$$
· Thus, if sites S_1 and S_2 are connected:
$$d_s (S_1 S y_2), = d(S_1 S_2) + d(S_1 S_2) - d(S_1 S_2).$$
Similarly, if sites S_1, S_2 and S_3 are connected:
$$d_s(S_1 S_2 S_3) - d(S_1 S_y) + d(S_x S_2) + d(S_2 S_y) + d(S_x S_3) - d(S_1 S_2) - d(S_2 S_3)$$
$$= \{d(S_1 S_y) + d(S_x S_2) - d(S_1 S_2)\} + \{d(S_2 S_y) + d(S_x S_3) - d(S_2 S_3)\}$$
$$= d_s(S_1 S_2) + d_s(S_2 + S_3).$$

Hence, if all the n sites could be connected, the problem is reduced to the travelling salesman problem. In this case the distance saved is given as:
$$d_s(S_1 S_2 \ldots S_n) = d_s(S_1 S_2) + d_s(S_2 S_3) + \ldots + d_s(S_{n-1} S_n).$$

$$= \sum_{r=2}^{n} d_s \left(S_{r-1} S_r \right)$$

It should, however, be noted that the order in which the sites are connected is important. The distance saved in linking site S_i to site S_j is not equal to the distance saved in joining site S_j to site S_i. For example, consider the linking of site S_k to site S_m. Distance saved is given as:
$$d_s(S_k S_m) = d(S_x S_m) + d(S_k S_y) - d(S_k S_m),$$
which is clearly not equal the distance saved when site S_m is linked to site S_k in spite of the fact that $d(S_m S_k) = d(S_k S_m)$.

Thus, when designing paths for vehicles the order in which the sites are linked is very important since $d_s(S_i S_j) \neq d_s (S_j S_i) \ \forall i \neq j$.

2. THE ALGORITHM FOR THE REFUSE COLLECTION PROBLEM

a) Definitions and Assumptions

The following terms, which are necessary to understand the proposed algorithm, are defined here:
1) The *servicing of a collection site* means the collection of refuse from a residential suburb or a commercial/industrial site. For our purpose we call them as collection sites.
2) A *trip* of a vehicle is the dispatch of the vehicle from the depot to one or more collection site(s) and when full/nearly-full goes to the dump for disposal. After the disposal it returns to the depot which completes one trip. However, if it is assigned to go to some other collection site(s) from the dump, it is said to perform a second trip. Thus, a trip is the number of visits of a vehicle to the dump site after collection before it returns to the depot.

3) A *task* is the servicing of a set of sites in a specific order selected from the group of sites, which are due for service on a particular day.

4) The *order* of a collection site gives the position of the site within a set of sites, which have been arranged in accordance to some specific rule. Here the collection sites are arranged in decreasing distance starting from the furthest site to the nearest site with reference to the dump. They are then positioned in ascending *order* starting with the furthest site as site 1 and ending with the nearest site to the dump as site *m*. The symbol *"namesite (j)"* is used to represent a site where name denotes the physical location of the site and *(j)* is the order. For any given site, the name will not change, but the order changes after each task.

5) *Assume* that a certain day *i*, the situation is:

Number of sites to service is m_i

There are three types of vehicles available, *A*, *B* and *C*:

Capacity of type *A*   = 1000 bins (household refuse bins)
Capacity of type *B*   = 625 bins
Capacity of type *C*   = 525 bins

Due to constant breakdowns of vehicles, the number of available vehicles for each type may change every day. It is assumed that the total number of available vehicles on day *i* is denoted by v_i, where,

number of available vehicles of type $A = a_i \geq 0$
number of available vehicles of type $B = b_i \geq 0$
number of available vehicles of type $C = c_i \geq 0$
and $v_i = a_i + b_i + c_i \geq 0$.

6) *Demand* at each site *j* varies and is known. At site *j*, the demand is denoted by q_j $(j = 1, 2, ..., m_i)$. The demand at any site is such that it can be serviced by any vehicle available on the day in a single trip (*i.e*, the demand at any site is not more than 525 bins). Moreover, one vehicle can serve l_i number of sites, where l_i is an integer ≥ 1 provided the demand of these sites satisfy the capacity constraint of the vehicle.

It is to be noted that although the sites that require service are known in advance, the operation is dynamic in the sense that the number of vehicles available every day differs. Also, though the proposed algorithm to follow consider only three types of vehicles to address the problem faced in the City of Bulawayo, the concept is general and can be accommodated to any number of vehicles of varying capacities.

b) The Problem

Given v_i and m_i, minimize the number of trips and find for each trip the associated task. All sites need to be serviced and total distance covered is kept as low as possible.

c) The Proposed Algorithm

Step 1
Initialize the actual position on day i: m_i—Arrange the sites in order of decreasing distance, starting with the furthest site in the order *namesite* $\alpha(1)$, *namesite* $\beta(2)$, ..., *namesite* $\delta(m_i)$, $m_i > 0$. Also, find the values of a_i, b_i, c_i, v_i, and q_j for $j = 1, 2, ..., m_i$.

Let $\xi = \{namesite\ (j)\}$, $j = 1,2, ..., m$; be the set of sites arranged in decreasing distance from the dump such that *namesite* $\alpha(1) >$ *namesite* $\beta(2) > ... >$ *namesite* $\delta(m_i)$.

Set $n(\xi) = m_i$.

$K = 0$, where K represents the counter for the total number of trips and let the counters for vehicle types A, B, and C be denoted by KA, KB and KC respectively.

Set $KA = 0$, $KB = 0$ and $KC = 0$.

Let the number of remaining sites to be serviced after trip K be (rk), i.e., $(r0) = m_i$.

Let $\xi_{(rk)}$ be the set consisting of the remaining sites, arranged from furthest site to nearest site with reference to the dump in decreasing order of distance, after some sites have been allocated, and $\xi_{(sk)}$ be the set of the allocated sites after K trips.

Then $\xi_{(rk)} = \{\xi\}$ and $\xi_{(sk)} = \phi$

Step 2
If $a_i = 0$, go to Step 4.
If $a_i > 0$:
Set $KA = KA + 1$ and dispatch the largest truck (type A) to the site 1, furthest form the dump. Starting from there the vehicle services this site and if there is still room (capacity) left, the same vehicle is used to service the site 2 on its way to the dump.

If the demand q_2 (for site 2) cannot be satisfied by the remaining capacity of the truck, we go on trying the next site(s) for which the demand can be fully satisfied in the remaining capacity, and when the truck is full/nearly full or no other site can be serviced in the remaining capacity of the truck, it goes to the dump. This completes a trip for a type A vehicle.

Set $K = K + 1$ and $a_i = a_i - 1$.

Re-order the remaining sites furthest to nearest from the dump in the decreasing distance: *namesite* $\alpha(1)$, *namesite* $\beta(2)$, ..., *namesite* $\delta(rk)$, where $(rk) < n(\xi)$.

Set $\xi_{(rk)} = \{namesite\ (j)\}$, $j = 1, 2, ..., (rk)$.

Denote the sites which have been serviced by $\xi_{(sk)}$.

Set $\xi_{(sk)} = \{\xi - \xi_{(rk)}\}$.

Set $\xi = \{\xi - \xi_{(sk)}\}$, $n(\xi) = n(\xi - \xi_{(rk)}) = (rk)$

Output:

"The trip number" K "for the vehicle type A serves the sites" $\xi_{(sk)}$.

"The task" KA "comprised S_x," $\xi_{(sk)}$", S_y, S_x".

Go to Step 3.

Step 3

If $n(\xi) > 0$, go to Step 2.

If $n(\xi) = 0$, go to Step 8.

Step 4

If $b_i = 0$, go to Step 6.

If $b_i > 0$:

Set $KB = KB + 1$ and dispatch the type B vehicles to the site 1, furthest from the dump. Starting from there this vehicle services this site and if there is still room left, the same truck services the next site, provided the demand and the capacity constraint of the vehicle is satisfied. If the second site does not satisfy the capacity constraint of the vehicle, the next site(s) is/are then considered until the vehicle is full/nearly full. When the vehicle has reached capacity/near capacity or when no other site can be serviced in the remaining capacity of the truck, it goes to the dump. This completes a trip of type B vehicle.

Set $K = K + 1$ and $b_i = b_i - 1$.

The remaining sites are then re-ordered: *namesite* $\alpha(1)$, *namesite* $\beta(2)$, ..., *namesite* $\delta(rk)$, where $(rk) < n(\xi)$,

Set $\xi_{(rk)} = \{namesite\ (j)\}$, $j = 1, 2, ..., (rk)$.

If $\xi_{(sk)}$ denotes the set of sites serviced, then $\xi_{(sk)} = \{\xi - \xi_{(rk)}\}$.

Set $\xi = \{\xi - \xi_{(sk)}\}$ and $n(\xi) = n(\xi - \xi_{(sk)}) = (rk)$.

Output:

"The trip number" K "for the vehicle type B serves the sites " $\xi_{(sk)}$. "The task" KA "comprised the elements in the set $\{S_x,$ " $\xi_{(sk)}$", S_y, $S_x\}$.

Go to Step 5.

Step 5

If $n(\xi) > 0$, go to Step 4.

If $n(\xi) = 0$, go to Step 8.

Step 6

If $c_i = 0$, go to Step 9.

If $c_i > 0$:

Set $KC = KC + 1$ and dispatch a type C vehicle to the furthest site 1. Starting from there the vehicle services this site and if the vehicle has not reached its capacity, the same vehicle is used to service the next site provided the demand and the capacity constraint of the truck is satisfied. If this site does not satisfy the capacity constraint, the successive site(s) is then considered until the capacity of the truck has been reached/nearly-reached. When the vehicle is full/nearly full or no other site can satisfy the remaining capacity of the truck, it goes to the dump. This completes a trip for a type C vehicle.

Set $K = K + 1$ and $c_i - c_i - 1$.

The remaining sites are then re-ordered in decreasing distance, starting from furthest site from the dump as *namesite* α (1), *namesite* β (2), ..., *namesite (rk)* where $(rk) < n(\xi)$.

Set $\xi_{(rk)} = namesite$ (j)], j = 1,2, ... (rk).

Set $\xi_{(sk)} = \{\xi - \xi_{(rk)}\}$

Set $\xi = \{\xi - \xi_{(sk)}\}$ and $n(\xi) = n(\xi - \xi_{(sk)}) = (rk)$.

Output:

"The trip number" K "for the vehicle type C serves the sites" $\xi_{(sk)}$.

"The task" KC "comprised the elements in the set $\{S_x, \xi_{(sk)}$ ", Sy Sx}".

Go to Step 7.

Step 7

If $n(\xi) > 0$, go to Step 6.

If $n(\xi) = 0$, go to Step 8.

Step 8 Stop. Print "End of allocation".

Step 9 Stop. Print "Short of vehicles".

3. NUMERICAL ILLUSTRATION

a) Problem Formation and Given Data

Consider a case where the number of sites required to be serviced is n. Let $n = 12$. Table 1 shows the distance between pair of sites in kilometres. The

Table 1. The demand at each site and pairwise distances

Demand in bins	Distance between	S_1	S_2	S_3	S_4	S_5	S_6	S_7	S_8	S_9	S_{10}	S_{11}	S_{12}
						sites							
240	S_1	0	5	12	22	21	24	31	35	37	41	49	51
340	S_2		0	7	17	16	23	26	30	36	36	44	46
300	S_3			0	10	21	30	27	37	43	31	37	39
280	S_4				0	19	28	25	35	41	29	31	29
340	S_5					0	9	10	16	22	20	28	30
250	S_6						0	7	11	13	17	25	27
240	S_7							0	10	16	10	18	20
340	S_8								0	6	6	14	16
510	S_9									0	12	12	20
440	S_{10}										0	8	10
525	S_{11}											0	10
525	S_{12}												0

Table 2. Fleet, available vehicles, capacity and their allocations.

Trucks (capacity in bins)	525	625	10 000
Total Fleet	17	11	13
Available	10	6	4
	(C_i)	(b_i)	(a_i)

demand at each site, which is assumed to be known in advance, is shown in terms of number of bins.

From Table 2, we gather that the fleet consists of 41 trucks of various capacities. Assume that on a particular day, due to constant breakdowns and lack of spare parts, only 20 trucks are available for servicing the sites.

Our aim is to minimize the number of vehicles to be allocated to the collection sites and find the task allocated to each vehicle. The distance of each site (S_i) from the Depot (S_x) and to the Dump (S_y) have also been computed using the most practical routes and arranged in increasing order from the dump node as showing Table 3.

Table 3. Distance of each site from Depot and to Dump

Sites (S_i)	Depot (S_x)	Dupm (S_y)
S_1	8	9
S_2	10	14
S_3	18	21
S_4	20	23
S_5	19	22
S_6	22	25
S_7	28	32
S_8	31	36
S_9	33	38
S_{10}	38	42
S_{11}	43	50
S_{12}	44	52

Using the method that was discussed earlier in Section l, we calculated the distances saved when linking sites to one another, these are given in Table 4.

b) Vehicle Deployment

The sites (starting with the furthest site from the dump) that yield the maximum distance saved are then linked until their combined demands do not exceed fire capacity of the truck to be allocated.

Table 4. Showing the distance saved when linking the sites

	T_0	S_1	S_2	S_3	S_4	S_5	S_6	S_7	S_8	S_9	S_{10}	S_{11}	S_{12}
LINK													
S_1		–	14	15	07	07	07	06	05	05	06	03	02
S_2		17^e	–	25	17	17	13	16	15	11	16	13	12
S_3		17	24	–	31	19	13	22	15	11	28	27	26
S_4		09	16	31^d	–	23	17	26	19	15	32	35	38
S_5		09	16	19	23^d	–	35	40	37	33	40	37	36
S_6		09	12	13	17	35	–	46	45	45	46	43	42
S_7		09	16	23	27	41	47^c	–	53	49	60	57	56
S_8		09	16	17	21	39	47	54	–	63	68	65	64
S_9		09	12	13	17	35	47	50^c	63	–	64	69	62
S_{10}		09	16	29	33	41	47	60	67	63	–	77	76
S_{11}		09	16	31	39	41	47	60	67^b	71	80	–	84
S_{12}		09	16	31	43	41	47	60	67	65	80^a	85	–

For example, start with site S_{12}. Linking S_{12} with S_{11} would result in a maximum "saving" of 85 units. But this is not possible because the combined demand (1050) would exceed the maximum capacity of the biggest truck (1000 bins) available. Hence, we look for the next promising site, S_{10}. Linking S_{12} and S_{10} gives us a "saving" of 80 units and the combined load is 965 bins < 1000 bins. Thus, the first trip has been designed, which could be labelled as trip 1 of vehicle type A.

In other words, $K = 1$ and $KA = 1$. Hence, the task associated with trip 1 is: $S_x \rightarrow S_{12} \rightarrow S_{10} \rightarrow S_y \rightarrow S_x$ with a total of 965 bins collected. The distance saved is 80 units.

Using the same approach, the other trips could be found and labelled accordingly.

Assign the 2nd trip to a vehicle of type A. Thus, the task associated with trip 2 is: $S_x \rightarrow S_{11} \rightarrow S_8 \rightarrow S_y \rightarrow S_x$. The total number of bins collected is 865. Distance saved is 67 units. Hence, $K = 2$ and $KA = 2$.

Assign the 3rd trip to a vehicle of type A. The associated task for trip 3 is: $S_x \rightarrow S_9 \rightarrow S_7 \rightarrow S_6 \rightarrow S_y \rightarrow S_x$. The total number of bins collected is 1000 and distance saved is 97 units. Hence, $K = 3$ and $KA = 3$.

Assign the 4th trip to a vehicle of type A. The associated task for trip 4 is: $S_x \rightarrow S_5 \rightarrow S_4 \rightarrow S_3 \rightarrow S_y \rightarrow S_x$. The total number of bins collected is 920 and distance saved is 54 units. Hence, $K = 4$ and $KA = 4$.

Assign the 5th trip to a vehicle of type B (capacity 625 bins). The associated task for trip 5 is: $S_x \rightarrow S_2 \rightarrow S_1 \rightarrow S_y \rightarrow S_x$, where the total number of bins collected is 580 and distance saved is 17 units. Hence, $K = 5$, $KA = 4$, $KB = 1$ and $KC = 0$.

Table 5. An allocation plan

Trucks (capacity in bins)	525	625	1000
Available	10	6	4
Allocated	0	1	4
	(C_i)	(b_i)	(a_i)

Although the optimality of the above trips with respect to distance covered is not guaranteed, the results generated are good and the method used is effective and simple.

Table 5 shows an allocation plan where only five trucks are allocated to service the 12 collection sites. In a recent study earned out by Bappoo and Kumar (1996) on the garbage collection system in the City of Bulawayo, it was found that each vehicle was assigned to a site for refuse collection.

Hence, in this case, 12 vehicles would have been used to service the 12 sites. Therefore, the proposed algorithm has greatly reduced the number of vehicles required to service the same number of sites. Moreover, in addition a total distance of 315 units is saved in this process (see the Appendix), the associated manpower to the vehicles has also been drastically reduced.

FURTHER INVESTIGATIONS

a) The Protean Cases

In developing countries like Zimbabwe, where spare parts are scarce and funds are limited, there is always likely to be a shortage of vehicles. The problem of allocating tasks to trucks is compounded in complexity by the fact that the fleet of vehicles is old and the number of trucks which are operational may not be sufficient to serve all the sites. This concept of disparate demand and supply has been discussed by Kumar (1995) as a protean system. Kumar and Bappoo (1995) have discussed the shortest routes in a protean network. In all these cases, it is vital to meet the task (demand). Garbage collection forms a protean system since garbage must be cleared in spite of the fact that trucks may or may not be available. Kalyan and Kumar (1987) have considered an allocation problem of similar nature where the number of jobs to be completed exceeds the number of workers available for the jobs. Here, due to constant breakdown of vehicles it has been observed that very often the number of vehicles available may not be sufficient to service all the sites in one trip per day. Hence, a need will arise when some vehicles will have to do more than one trip per day.

Moreover, since the number of vehicles available each day differs, the vehicle deployment to sites can be classified as a protean system. When the number of vehicles is more than the required number of trips needed to serve that day, no problem is encountered in deploying the vehicles to the collection sites. The solution can be found by using the heuristic approach as shown in Section 3(b). An interesting problem arises when the number of vehicles is less than the number of trips needed to serve all sites on a particular day. This problem may be tackled through the introduction of overtime or double shifts for some vehicles, which will automatically increase the operational costs in the collection of garbage.

The model presented in Section 2 needs to be modified to consider the protean cases which will be dealt in a later publication. We assumed that each vehicle could serve a site in a single trip. But it is possible that the sites may be larger or may expand over the time, thus increasing the demand. Hence, only bigger trucks with greater capacity can service such sites in a single trip, or alternatively some vehicles will have to make more than one trip to service the same sites.

This problem is compounded with so much complexity if new vehicles are not purchased to replace the old ones or the demand of a particular site does not satisfy the capacity of the vehicle. In such a case the vehicle will have to do more than a trip to service a single site.

ACKNOWLEDGEMENTS

The authors are grateful to the Research Board, National University of Science and Technology for their support. R. Bappoo wishes to thank the Department of Applied Mathematics for providing the necessary facilities.

REFERENCES

Bappoo, R. and S. Kumar. 1996. Vehicle Routing for Municipal Waste Collection: A Case Study, A report submitted to the Bulawayo City Council. Paper also presented at the Fifth Symposium on Science and Technology, *Towards Capacity Building in Science and Technology*, Research Council of Zimbabwe, Harare, pp. 37, Abstract 3.24.

Beltrami, E. and L. Bodin. 1974. Networks and vehicle routing and municipal waste collection, *Networks* 4(1): 65–94.

Bodin, L. and B. Golden. 1981. Classification of vehicle routing and scheduling. *Networks* 11: 97–108.

Bramel, J. and D. Simchi-Levi. 1995. A Location based heuristic for general routing problems. *Opns. Res.* 43 (4): 649–660.

Clark, G. and J.W. Wright. 1964. Scheduling of vehicles form a central Depot to a number of delivery points. *Oper. Res.* 12: 568–581.

Dantzig, G.B. and J.H. Ramser. 1959. The truck dispatching problem. *Management Sci.* 6: 80–91.

Golden, B.L., J.S. DeArmon and E.K. Baker. 1983. Computational experiments with algorithms for a class of routing problems. *Comput. & Ops. Res.* 10 (1): 47–59.

Kalyan, R. and S. Kumar. 1987. Modeling and analysis of augmenting systems—An allocation problem. In *ASOR '87*, Proc. of the 8th National Conference of the Australian Society of Operations Research, Melbourne, pp. 126–136.

Kumar, S. and R. Bappoo. 1995. Shortest routes in protean networks. In Proc. of the International symposium on Operations Research and Applications, 1995, Ding-Zhu Du, Xiang-Sun Zhang and Kan Cheng (eds.), *Lecture Notes Series in Operations Research* World Publishing Co., Beijing, China, pp. 228–237.

Kumar, S. 1995. *Optimization of protean systems: A review In APORS 1994*, M. Fushimi and K. Tone (eds.), World Scientific Publishers, Beijing, China, pp. 139–146.

Appendix

The total distance covered when 12 vehicles are allocated to service the 12 sites is 678. Table 3 has been used to compute this distance.

Tables 1 and 3 are used to find the distance covered when applying the proposed algorithm.

Trip 1:	44 + 10 + 42	= 96
Trip 2:	43 + 14 + 36	= 93
Trip 3:	33 + 16 + 7 + 25	= 81
Trip 4:	19 + 19 + 10 + 21	= 69
Trip 5:	10 + 5 + 9	= 24

Hence the total distance covered is 363 km.

Therefore, distance saved in following the proposed method is 678–363 = 315 km, which confirms the results as given in Table 4. Each saving has been marked corresponding to its trip by an alphabetical index.

Section 3

Country Case Studies of Asian Continent

9

Introduction to Waste Management in Asia

Vaneeta Kaur Grover and Velma I. Grover

411–981 Main St. West, Hamilton, On, L8S 1A8, Canada

The Asian continent extends from the Asian part of Siberia in the north to Indonesia in the south and from Turkey in the west to Japan in the east. Asia is a huge continent with different dynamics and diversities—with the highest mountains, landlocked countries, lakes, seas and diverse climates; cultures and traditions. It has the most populous countries of the world— China and India—so only with 23% of total land area it has 58% of the total world population. As a result of high population and growing economies— like five tigers—in this region there are many problems arising out of economic development, over-population and pollution. Although, the environmental concerns vary widely across the continent some of the issues gripping it are:

- Land degradation
- Deforestation
- Deterioration in water quality
- Degradation of marine and coastal regions
- Growth of mega-cities resulting in social, economic and urban problems including problems related to solid waste management.

This region is undergoing economic and industrial growth and therefore the environmental problems related to the handling of the waste and its impact has accelerated. Although, the impact of waste mismanagement is not to the scale of global warming or ozone depletion, it has adverse impacts like the spread of diseases (dengue, plague, etc.). So, this economic and industrial development, must be followed by adequate measures and education, to combat the most urgent environmental problems including the problems of waste management.

Environmental degradation is especially serious and evident in the mega cities and their surrounding metropolitan areas mainly because they have become overcrowded in the extreme, due to the influx of migrants from the rural areas to the metropolitan centres. Most of these cities have expanded their boundaries in an uncontrolled manner, migrants have settled on the outskirts of the cities, as well as within the city boundaries in vacant lots, abandoned buildings, and such areas. As a result of this migration the capacity of municipalities (human, financial etc.) to provide even the most basic services has greatly exceeded. Uncontrolled and improper management and disposal of municipal solid wastes and contaminated water sources are major threats to the public health and environmental quality in most of the Asian region.

DATA COLLECTION AND PLANNING

There is a lack of proper and reliable data in most parts of Asia, mainly because of the lack of locally available trained personnel. The external advice is mostly sought for the management of waste, but external advice is of little use unless the advisors are aware of the vast differences in the social, cultural, financial, and environmental conditions and in the waste characteristics between the Asian countries and those of the country from where the advisors came. Understanding the conditions requires the collection of certain key data as well as knowledge of the social, cultural, financial, and environmental conditions prior to the preparation of plan of action. The implementation of foreign technology does not help in these regions.

Some of the required basic data are quantity, composition, and characteristics of the waste generated. According to a UNEP report, around 700 million tons of solid waste is generated each year in Asia and among all Asian regions, East Asia, generates the largest share of municipal solid waste. According to the same report, its share may increase to 60% by the year 2000 because of the large population base and high economic growth rate. But the disposal of waste is given very low priority in many of these countries, according to a report of ESCAP, only 70 % of the waste, in urban areas, is collected and only 5% is treated. Table 1 gives the composition and total waste generated in several Asian cities (SWM for developing countries: Diaz *et al.* 1996)

In addition, for proper waste management, information is needed on current waste management practices such as storage, collection, final disposal, availability of equipment, maintenance procedures, availability of human resources, budget, and sources of revenue. The characterization of the waste is also an important element in the development of realistic and sustainable solid waste management programmes, because successful management and processing of wastes depends on the type and composition of the material. Waste generation rates and composition vary substantially among Asian countries as is apparent from the data in Table 1—in the percentages of paper; glass, plastics/rubber/leather, and ceramics/dust/

Table 1. Waste generated in Asian cities

Location	Putre-scibles	Paper	Metals	Glasses	Plastics rubber, leather	Tex-tiles	Ceram-ics, dust, stones, ashes	Others	Wt/ca p/day (g)
Bangalore, India*	75.2	1.5	0.1	0.2	0.9	3.1	19.0	–	400
Israel*	71.3	24.8	1.1	1.0	0.8	1.0	1.8	–	400
Manila, Philippines*	45.5	14.5	4.9	2.7	8.6	1.3	27.5	–	400
Seoul, Korea*	22.3	16.2	4.1	10.6	9.6	3.8	33.4	–	2000
China	28	2	1	1	1	–	66	1	–
Japan	57	25	1	3	10	–	–	4	–
Thailand**	39.2	13.6	1.9	1.1	14.5	4.8	–	24.9	–
Iraq**	68.6	10.2	2.3	2.4	3.9	3.8	5.5	3.3	
Hong Kong**	46.2	25.7	1.9	5.6	8.4	9.0	0.4	2.8	–
Abu Dhabi **	22.5	42.4	14.0	4.4	6.3	0.3	3.8	6.2	–
Lahore, Pakistan$	49	4	4	3	9	5	2	24	–
Sri Lanka (Residential) #	80	8	1	6	2	1	–	2	

* Diaz et al., 1996.
** Manipulated from data given in UNESCO module on Solid Waste Management.
$ Manipulated from data given by Madhusmita Moitra, Management of Solid Waste: Key to Sustainable Development.
#. Solid Waste Management in Sri Lanka: A Country Perspective by Dr. Ajith de Alwis (1999).

stones. It can be seen that the Japan, which is considered most developed among the countries listed in Table 1 has the highest quantity of waste paper (the higher the development the higher is the quantity of paper in the waste) next to Abu Dhabi. The high content of paper in Abu Dhabi might be because of the affluence in the country attributed to oil. Otherwise, the organic content (putrescibles) is maximum in the waste of all the Asian countries, which increases the moisture contents and bulk densities of waste.

COLLECTION, TRANSPORT OF WASTE

A variety of methods and equipment are used for the collection of wastes ranging from labor-intensive methods to fully mechanized ones. In the rural areas of most of the Asian countries the waste is either still carried on the heads, in handcarts, bicycles and ox carts. In urban areas there might be the use of handcarts for the transportation of waste from homes to the bins or collecting points. Otherwise trucks or the most modern vehicles are used for the collection and transportation of waste to the landfill or disposal sites.

In most countries of Asia, the compactor trucks, both of the rear and front loading variety, are becoming increasingly popular. This is not good because it has been observed that generally very little, compaction (if there is any at all) occurs in the vehicle. It can be attributed to the fact that the loose wastes have a high bulk density. Furthermore, as pointed out by Diaz *et al.*, (1996) some complex features and consequences are associated with the use of compaction vehicles, like the weight of the truck might exceed the bearing capacity of roadways, these vehicles are generally inaccessible to remote areas and narrow streets (and the streets are unpaved and narrow at many places), there is a requirement for adequate facilities and trained personnel to conduct complex repairs and preventive maintenance, particularly of the hydraulic system; and the need for readily-available source of spare parts to maintain the regularity of the collection service (the sense of preventive maintenance is lacking in most of these countries).

Another problem associated with the collection and transportation of the waste is that the collection routes are not firmly established. The common practice is to leave the decision for the route to the discretion of the driver, so either only partially loaded trucks arrive at the disposal site due to the inefficient routing or they take a longer route according to their convenience to work less or to do some other work on the way.

DISPOSAL AND PROCESSING

The waste is mostly dumped at the landfill sites—at most places, litter is left behind—the percentage of litter varies from place to place. Most Asian nations do not have sanitary landfill sites (and the sites are not even maintained), thus leading to leachet and other pollution problems. As the calorific value of the waste is not high at most places, incineration is not a good option. But according to a UNESCO module on Solid Waste Management in Japan, waste to energy plants account for 25% of solid waste disposal, Hong Kong and Singapore have incineration as a major disposal facility because of the scarcity of land. The high organic content in the waste has encouraged people to go in for composting—in India composting and vermi composting is being tried at many places. In fact, in Bhopal, 80% of the waste collected is composted by a private organization. Traditionally, in rural areas in Asia the organic waste is utilized in either feeding the animals, as compost (fertilizer) in the fields, or as a source of fuel for cooking.

In most Asian countries, there is an informal recycling system which works along with the formal system of municipal waste management system. There exists a class of people *kabaris*—who like vendors or hawkers go from home to home and buy recyclable material like paper, plastic, glass, and old clothes—thus preventing these things from going into dump sites. There are also rag pickers (scavengers) who collect waste from either the landfill sites or the collection points. At some places, like in Bangalore (India) 15% of the waste and in Karachi (Pakistan) 10% of the waste is collected by these people. For many people this is a source of livelihood. The scavengers and hawkers have connections with the middle-men who buy things from them and sell them to the industries. In Hong Kong, 90% of the paper collected and all the aluminium cans are exported.

In many Asian countries there is a trend for the privatization of the collection and disposal system. For example, in Manila, the collection and disposal system is 100% privatized. In Malaysia, it is soon going to be all privatized. The same trend has been observed in countries like Korea, Singapore and Indonesia. In India, at places the disposal has been privatized but it has been done only in some cities.

COSTS

The total cost spent on solid waste management is relatively difficult to calculate because the other activities like drain cleaning are also carried out from the same budget. But it has been estimated that anywhere between 10 and 40% of the municipal budget is utilized for solid waste management (Bhide, 1990). It has been seen that out of this budget, 70–80% is spent on collection, 20–30 % is spent on street sweeping and other overheads thus leaving little or no money for the disposal. This poor and inadequate management of Municipal Solid Waste Management leads to relatively high costs for the service provided (Table 2).

Since very few resources are devoted to final disposal, the disposal consists of throwing the wastes and spreading them upon the land in an uncontrolled fashion and without modern construction methods (e.g., small working face, bottom liner and leachate control system, and landfill gas control system).

CONCLUSION AND SUMMARY

In most of these nations the planners are becoming aware of the problems related to the mismanagement of solid waste and, thus, background studies

Table 2. Per capita expenditure on solid waste management in different countries:
Asnani, 1991

Country	$/Capita/Year
Bangladesh	0.36 in Dhaka and range 0.07-0.03 in other towns
India	2.38-4.15
Indonesia	0.84-1.9
Myanmar	0.75-8.4
Nepal	2.03 in Kathmandu and range 0.25-0.84
Sri Lanka	1.79 in Colombo, 0.9-1.1 in medium and 0.15 -0.25 in small towns
Thailand	1.6-5.35

and surveys are being conducted, and courses of action are being prepared. The prospects for success of these plans will depend largely on to what extent the plans are implemented and how best the methods selected are suited to the needs of the local conditions.

REFERENCES

Ajith de Alwis, 1999. Solid Waste Management in Sri Lanka: A Country Perspective (in this book)

Asnani, 1991. Supplementary background document, prepared for consultation on Solid Waste Management in South East Asia Region, New Delhi, World Health Organization.

Bhide, 1990, Regional Overview of Solid Waste in South-East Asia Region.

Diaz, F.L., G.M. Savage, L.L. Eggerth and C.G. Golueke, 1996. Solid Waste Management for Economically Developing Countries.

Madhusmita Moitra, 1996. Management of Solid Waste: Key to Sustainable Development in Jan - June 1996 (Vol. 3, No 3 & 4) Newsletter of Envis Centre on Human Settlements.

UNESCO, unpublished, Module on Solid Waste Management—A Module. Phelps, H.O. (Co-ordinator).

10

Solid Waste Management of Hong Kong: Its Evolution and What Developing Countries can Draw on

Chung Shan Shan

Environmental Engineering Unit, Civil & Structural Engineering Department,
Hong Kong Polytechnic University,
Hung Hom, Hong Kong

INTRODUCTION

Unlike the environmental problems associated with air and water pollution, the problems with solid waste disposal can be delayed and covered up for a longer period. For this reason, the priority of resource allocation to rectify the solid waste problem is generally lower than other more immediate and prominent problems. For example, the Clean Air Act of the UK was passed in 1956, but laws on solid waste management which was enveloped in the Environmental Protection Act, became law in 1990. Similarly, in the USA, the Clean Air Act was passed in 1970 with national ambient air quality standard set. But the Resources Conservation and Recycling Act was passed six years later and its main objective was to control hazardous waste.

Thus, it is not surprising to find that environmental impacts resulting from poor solid waste management are still unsolved problems in developed countries not to mention the developing countries. Furthermore, not until recently did policy makers realize that sound solid waste management should be formulated in the context of sound resources management. Such a delay in identifying the right policy direction shall not be repeated in the developing countries.

The evolution of solid waste management in the developed world provides valuable experience for others to draw on. It is the aim of this paper to use the solid waste management experience in Hong Kong to illustrate the mistakes that should be avoided by the policy makers in the developing countries.

HONG KONG AND ITS SOLID WASTE MANAGEMENT
MECHANISM

Solid waste management is an integral part of society. A clear understanding of the solid waste management problems of society involves knowing some of its socioeconomic features. The following short introduction will help to understand the subsequent discussion.

The Geography and the Economy

Although Hong Kong was once a British colony, it has always been a Chinese society and is now a special administrative region of the People's Republic of China. The per capita GDP of Hong Kong in 1997 was HK$ 204,092 (about US$26,166). Environmental spending accounted for about 2.9% of the total government spending in 1997. In other words, although Hong Kong is part of a developing country (i.e. the People's Republic of China), it is a high income economy.

With a terrestrial area of 1092km² and a population of 6.68 million in 1998, the average population density of Hong Kong is one of the highest in the world. The average population density of the old urban districts is about 26130/km² (Hong Kong, 1996).

Thus land which is indispensable in all sorts of development, is a premium in Hong Kong. Land prices and living indices have been increasing rapidly and have become driving factors for the high labor cost in Hong Kong. Owing to the rapid increase in population and development needs, the civil and construction industry has also been growing hand in hand with the land and property prices. As a result, a large quantity of construction waste has been generated and has become a major management problem.

Positive Non-intervention and the Environment

The governing philosophy of Hong Kong is "positive non-intervention". That is, the government generally will not interfere with the economy but will provide the infrastructure to enhance its economic development. The freedom given to the private sector was extended to environmental and pollution control as well.

Treating environmental protection seriously has been a very recent phenomenon compared with the long history of the British rule. Major pollution legislation did not appear until 1980. Before that, control on industrial and other forms of pollution was minimal and the regulatory duty was diffused over a number of government offices.

The Evolution of Waste Management Strategies

The earliest official environmental protection body, the Environmental Protection Unit (EPU) was established in 1977. The EPU was charged with coordinating various environmental protection activities which were carried out at that time by a dozen or so different government departments. However, during the last two decades, there has been considerable regrouping of environmental policy formulation duties within the government. In 1986, the Environmental Protection Department (EPD) was formed to take care of environmental control in Hong Kong. In 1998, the EPD had 1628 staff and an estimated total expenditure of HK$ 2244.6 million[1] (EPD, 1998).

The standard in waste management has been raised with the growth in the size of the environmental authority. In the past, landfill sites in Hong Kong has been built to simply bury waste. There were little environmental precautionary measures so to speak. Remedial measures to the related environmental problems did not become a pressing need until urban sprawl appeared and population centres encroached onto previously remote areas where the old landfills were built. Restoration work was therefore necessary to reduce the hazards of old landfills (especially landfill gas migration) to the nearby population (EPD, 1996a) . On the contrary, the three extant strategic landfills in Hong Kong are equipped with a double lining system, a landfill gas and leachate collection system. All leachate will also be treated on the site before it is discharged into the aquatic environment.

Before 1989, the waste treatment approach was more diversified compared to the existing one. There were once three incinerators in operation and landfills were counted on for the rest of the waste. However, the pollution from incinerators was found to have a increasing detrimental impact on the neighbourhood, partially due to urban sprawl and poor urban planning. This has led to the formulation of a landfilling oriented policy in 1989 (Planning, Environment and Lands Branch, 1989). According to the 1989 strategy, the incinerators will be phased out in stages and three large landfills with a total capacity of 135 million metric tons would be used to accommodate the solid waste in Hong Kong (EPD, 1996b). The shortcomings of such a strategy will be analysed later.

CONSTRUCTION WASTE IN HONG KONG

Construction waste has traditionally been transferred either to public dumps (usually land reclamation sites) or to landfills. In the early 1990s, there was a decline in public dumping space and most of the construction waste was then diverted to the landfills for disposal. At the peak, up to

15,000 tpd of construction waste was disposed of in landfills, almost twice the quantity of MSW of that year (Figs.1 and 2). Throughout the 1990s, construction waste has been the largest category of waste disposed in landfills. Nevertheless landfilling is not a sensible way to dispose construction waste. Figure 1 shows the amount of construction waste generated in Hong Kong from 1992-1998.

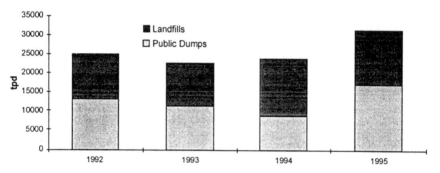

Fig. 1. Construction waste disposal in Hong Kong (1992-1995)
(Sources: EPD, 1993-1996a)

Owing to the geographic characteristics, reclamation is often carried out to obtain the land required for development. Rather than dredging marine sands from the undisturbed seabed, it would be more environmentally sensible to use the earthwork material from construction waste for the purpose of reclamation. The physical potential for such use of construction waste is enormous, as about 80% of the demolition waste and up to 96% of the renovation waste can be used in reclamation or fill which is demanded in large quantity in Hong Kong (HK Polytechnic and HK Construction Association Ltd., 1993). This has, however, not been extensively carried out due to the lack of incentives for the construction contractors to separate the non-fill portion, namely, wood plastic and metals from construction waste.

In order to discourage abuse of landfill space, a landfill charge that reflects the environmental and financial costs of landfilling should be introduced. According to an earlier study, the full cost of landfilling ton of waste in Hong Kong is around HK$1000/ton. This figure does not include the cost of transporting waste to landfills (Chung, 1996).

Landfills charges also provide incentives for the construction industry to adopt more low-waste construction techniques such as reusing aggregates in parking lots and foundations, using pre-cast components, metal scaffolding, formwork and hoarding. It also encourages the construction contractors to do on site waste separation and to minimize the use of landfills. Nevertheless, waste disposal is still free in Hong Kong. Having said that, the government did propose to introduce a landfill charge for industrial, commercial and construction wastes at a rate of HK$86/ton which is, however, well below the full cost of landfilling.

MUNICIPAL SOLID WASTE IN HONG KONG

Municipal solid waste (MSW) is defined in Hong Kong as consisting of domestic, industrial and commercial waste. Over the past 10 years, the amount of MSW delivered for disposal in Hong Kong has been found to increase steadily largely due to the growth in population and economic activities with the exception of in 1994 and 1995 when a halt and even a decrease in MSW was experienced. There are two major reasons for the halt in MSW growth:
1. There is a reduction in industrial waste generation in Hong Kong due largely to the sectoral change in the economy, and
2. There is increasing interest in voluntary recycling schemes.

Figure 2 shows the waste generation rates. Total waste generation stood at about 6.02 million tons in 1998 (Ref. EPD, 1999b). About 33% (roughly 1.56 million tons) of this was recovered for export or for local recycling (EPD, 1999a). The waste recovery rate compares favourably to many developed countries. Nevertheless, what was left behind, namely 1.31 kg/capita/day (as in 1998) was still a burden for Hong Kong in the long term. In waste management, the availability of facilities to dispose the waste, regardless of the generation rate, is an important factor to consider. Hong Kong, being a small city, is particularly vulnerable to the lack of land for waste disposal and yet landfilling has been the major method to dispose solid waste until the present. Figure 3 compares the MSW pressure on each km^2 of land for Hong Kong and other nations. It shows that the average waste required for disposal in Hong Kong is very high for each unit of land.

Such intangible pressure on Hong Kong land has now emerged as a rapid shortening of landfill lives from the originally planned 30-40 years to the present 16 years (Stokoe, 1996). Faced with such pressure from waste management, a new waste management strategy is urgently needed in Hong Kong (see the Planning, Environment and Lands Branch, 1997). Whatever are included in the Plan, the basic solutions are already well-known: (i) to promote waste avoidance and reduction, (ii) to increase waste recovery, and (iii) to explore other waste disposal methods. In the following sections the current status of waste reduction, waste recycling and alternative waste disposal methods in Hong Kong are discussed.

WASTE REDUCTION IN HONG KONG

In the past, there has been a lack of emphasis on waste avoidance and

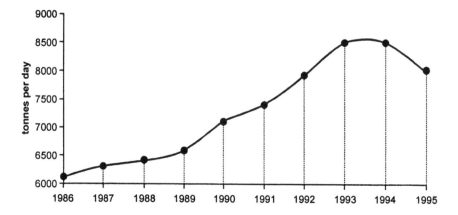

Fig. 2. MSW delivered for disposal in HK from 1986-1998
(Sources: EPD, 1986, 1988, 1993-1996a, EPD, 1999b))

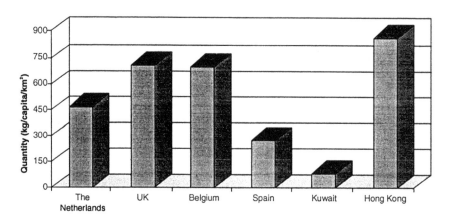

Fig. 3. A comparison of MSW pressure by nations (source: author)

reduction. There is no law and on incentive to encourage waste avoidance or reduction with the exception of chemical waste which is required by law to be disposed properly, coupled with a charging system.

Waste reduction activities in Hong Kong are voluntary, and most of them are ephemeral or *ad hoc*. Organized waste reduction activity is a recent phenomenon. In 1995, a "Use Less Plastic Bags Campaign" was organized by the government and the retail industry and managed to attract some large retail outlets to participate. It aimed to reduce 10% plastic consumption for the first year. Over 35 million plastic bags were reduced in the first year and some retailers reduced 33% of their plastic bag consumption (EPD, 1996a). Although such an achievement is encouraging, it is by no means sufficient and a lot more has to be done for the concept of resource conservation to be extensively practised.

WASTE RECOVERY IN HONG KONG

Contrary to a community-based waste recycling programme in the developed countries, waste recovery is a business in Hong Kong. The waste recovery industry can broadly be divided into three tiers although there are minor variations in specific material collection network. The interrelations of the various parties in the waste traders are depicted in Fig. 4 The waste recyclers turn the waste material into new or same nature products and sell them to the users of the recycled products. In Hong Kong, some waste recyclers not only reprocess the waste materials but also directly export the materials overseas.

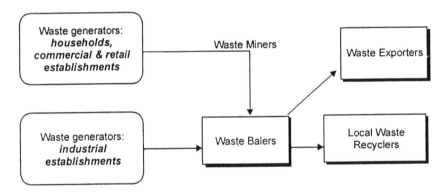

Fig. 4. A simplified diagrammatic illustration of the waste recovery system in Hong Kong

Scavengers or Waste Miners
The job of sorting marketable waste materials from household waste, public cleansing and commercial waste is left to waste removers, building cleaner, haulers and to a lesser extent, street scavengers. For the sake of simplicity, they are collectively called waste miners. These waste miners collect and sort the waste materials into different grades and categories. The materials they retrieve include aluminium cans, copper (in electrical cables), newspapers, cardboard, metals, clear glass bottles, rags, old white and brown goods, etc. Scavenging of marketable waste may occur very close to the point of waste generation such as within residential, commercial and industrial buildings, or further down, at municipal refuse collection points and even on the street, such as picking recyclables from public rubbish bins. The scavenging activities may sometimes cause nuisance or

inconveniences to the passerbys. Complaints on the scavenging activities are also received by the Municipal Councils as the scavengers mess up the site from time to time (Leung, 1995)

Waste miners sort the waste into types and grades and sell them to the waste traders in the next tier—the waste balers. Waste retrieval at this level is informal. This means that the activities and the work force are unregistered and generally unregulated. However, according to a public cleansing law, all the waste discarded in municipal collection points are the property of the Municipal Councils. Thus, strictly speaking, waste picking from these places are illegal. Nevertheless, this is seldom enforced.

Wastebalers

Waste miners sell the marketable waste to waste balers. Waste balers buy waste materials from the waste miners, further sort the waste and then compact them into transportable bales. Although they usually engage in the buying and selling of a variety of wastes with waste paper as the chief material, some of them are found to be specialized in one or two types of waste only, usually paper and aluminium cans or assorted metals. Waste balers are usually small-scale businesses with less than 10 employees.

Waste balers with two or less workers depend solely on waste miners working in the neighbourhood for waste supply. Their transaction output can be as little as 50 tons or less per month. The large waste balers may handle 200-400 tons of secondary materials per month. Larger waste balers may source from industrial waste themselves or collect waste in quantity from other neighbourhoods (usually from housing estates, schools and institutions) and haulers. Those relying on industrial waste supply usually have contractual agreements with large recyclables generators, such as printing or plastic factories.

Waste Reprocessors and Waste Exporters

The baled waste will be sold either to the local waste reprocessors or to exporters. Waste exporters do not usually deal directly with individual waste miners although exceptions could be found. Exportation is a more risky business than trading in local markets. Waste exporters have to bear the risk of lost bales and product damage during shipment as well as higher potential loss if the recyclers are dissatisfied with the quality of the secondary materials.

Waste exporters usually have extensive connections with the processing industries in other countries. Some of them are sourcing only for their affiliated plants overseas. Some exporters will also sell the baled secondary materials to local reprocessors if prices are attractive.

Recyclers on waste paper, clean plastic scrap, ferrous, non-ferrous metals and wood can be found in Hong Kong. Some plastic reprocessors import plastic waste rather than rely on indigenous plastic scrap but most of them rely on both sources to ensure an adequate supply of quality raw materials.

Two non-MSW recycling operations, oil and car tyre retreading, are found in Hong Kong too.

Despite the high waste recovery rate, there is no room for optimism and complacency as: (1) even better recycling achievements are needed to mitigate the waste pressure, and (2) the waste recovery sector in Hong Kong is facing a lot of difficulties that would adversely affect the long-term development of this industry.

Difficulties of the Waste Recovery Sector

THE SUSCEPTIBILITY OF THE WASTE COLLECTION SECTOR TO PRICE DECLINES
This is found to be the case especially for waste miners and balers. From the experience of a decline in world prices during late 1992 to early 1994, under the profit-making motives, the sourcing of indigenous waste would be reduced at the expense of other more profitable activities such as waste material re-export and the collection of waste will also be restricted to those with more stable demand, such as metals, aluminium and old corrugated cardboard. At that time, the larger waste balers cut waste paper collection by 40-50% by giving up the low value mixed paper. Consequently, waste paper export has dropped by 23% in 1993 over the previous year. Secondary material prices (mainly waste paper) dropped again in 1997 and 1998. The immediate effect is the closing down of the largest paper recycles in Hong Kong (Hong Kong Economic Times, 1998). Longer term effects are to be seen.

To an industry where the profit margin is low and the operating cost is rising steadily, such fluctuations in their revenue lead to an opportunistic way of operation which means that capital investment will be minimal and that long-term business connections are not important to them. In short, high operation cost and the uncertain business prospect are the two main reasons to hinder the expansion of the waste recovery sector.

Floor price guarantee is a general measure used in other countries to support recycling for instances, in Japan and France. The floor price subsidy should be based on the cost of the most efficient waste baler(s) and be given to licensed waste balers upon waste transaction documents (such as receipts). Other feasible economic measures include tax allowance, grants and soft loans (Chung, 1996). However in the newly released Waste Reduction Frame Work Plan, the favoured proposal in preserving the waste recovery sector is to grant land in affordable cost but in a short 2-7 years times which cannot solve the real problem confronting this sector.

LOW ADAPTABILITY OF THE WASTE COLLECTION SECTOR
Most waste balers and the small waste recyclers are poorly informed of the world secondary material market situation. For instance, most waste balers did not know that the booming of waste recycling activities in Europe was causing the price depression in 1992-1994. Early and accurate trade information would have enabled them to take appropriate remedial

measures and minimize economic losses, such as by increasing trade with local buyers instead of seeking for export, decrease the collection of low-valued recyclables and collect more higher priced secondary materials, such as ferrous metals.

Owing to the financial constraints and the waste balers'education levels, it is unlikely that they can do thorough market research. Nevertheless, the government Trade Department can disseminate local world secondary materials market information free to the waste traders, the waste balers in particular, in the from of newsletter and market reports.

Need to Raise the Technology Employed by the Small Waste Recyclers
Major differences exist in the recycling operation and the perceived business prospects of the small and large recycling establishments. Small recyclers generally just modify the backward technologies to perform their recycling operations. At the same time, with a weak financial background, they usually have a very short-term aim such as to maintain a living. As a result, the recycling equipment is primitive and inefficient. Environmental precautionary measures are usually ignored, although some recycling operations particularly those using solid fuels, are covered by the air pollution control laws [2]. Furthermore, small plant owners have only limited access to technical information and even if they are aware of better recycling alternatives, they lack the required capital to invest.

In this regard, to adopt the positive non-intervention approach to the recycling business will mean allowing the recycling process to be carried out in a less environmentally beneficial form. On the other hand, stricter pollution control and enforcement will probably drive the small recyclers out of business. In view of the environmental gains form recycling, it will be justifiable to help upgrading the small recycling operation by using public money (Chung, 1996). At the same time, lessons can be drawn on the US experience by setting up recycling zones where cheaper land, better infrastructure and soft loans are offered to encourage more and better recycling operations. Improvement on recycling technology can also increase the marketability of the recycled products.

A case to be referred is the Californian example to promote economic development and resources conservation in one go. Recycling market development zones are set up to attract small investment in the recycling business and to provide employment. This policy has been found to work well (Kuh and Mang, 1994) Since the service and financial sectors have been dominating our economy, there is an urge to revitalize the industrial sector in Hong Kong as a risk-aversion move. Developing clean production and secondary material-based industries are sensible moves to blend environmental improvement into industrial development.

UNDER REWARD TO THE WASTE COLLECTORS
The waste recovery industry is not only contributing to the GDP of Hong

Kong but it also helps to avoid substantial negative environmental externalities to the society. Further, by recovering the recyclables for the MSW streams, savings in waste transport, landfilling and incineration costs are realized. However, the waste recovery industry is not fully rewarded for their enhanced environmental gains.

There is a need to formally recognize their contribution to the society. In this regard, the establishment of a recycling system in which waste recoverers are rewarded with the social gains associated is most appropriate.

UNNECESSARY CASCADING OF PRIME GRADE SECONDARY MATERIALS
The outputs of the local recyclers are usually low specification products: raw materials for shoe sole, reinforcing steel bars and iron cast manhole cover. Yet, a lot of prime grade industrial scrap is used in this process. It means that the recycling sector in Hong Kong has not fully exploited the material relinking potential of the recyclables sourced and an unnecessary cascading of the secondary materials has resulted.

Sophisticated recycling technologies and better purification generally results in less cascading. But as already mentioned the local recyclers are either complacent with their existing knowhows or are not capable of further investing in better technologies. Again there is a role for the government to enhance and offer affordable technology transfer in Hong Kong. Often financial assistance and tax breaks are required to help jump-start companies with emerging technologies to become economically-viable technologies. A stronger local recycling sector will also decrease the vulnerability of the waste recovery sector to the unstable world market.

Alternative Waste Disposal Methods

Incineration
In the waste management history of Hong Kong, landfilling has been the major method to dispose of MSW although there were once three incinerators capable of handling 2550 tpd (PELB, 1989). The last one was closed in 1997. Landfilling is the only waste disposal method in Hong Kong before any incinerator is built.

The choice to use landfilling rather than incineration as the main disposal method in the 1989 waste diposal plan was grounded mainly on two reasons:
(1) Pollution from the older generation of MSW incinerators and their close proximity to residential areas had made the incinerator a very unpopular facility in Hong Kong.
(2) Incineration is generally financially more expensive than landfilling.
Yet these two arguments are no longer convincing in the present context because the new generation of incinerators is not necessarily an

environmental burden. A previous study shows that waste to energy (WTE) is superior to landfilling as a waste disposal technique even taking into account its higher financial cost (Chung, 1996). In particular, incineration has the advantage of significantly reducing the volume of MSW. It is therefore not surprising to find that the government is reconsidering the incineration option in the proposed waste reduction plan (Planning, Environment and Lands Branch, 1997).

Composting

Contrary to other recycling processes, direct government involvement is found in composting. The government-run composting plant is the Sha Ling Composting Plant (SLCP). The composting process takes 51 days to complete and currently only pig waste is composted in the government composting plant. The SLCP has a capacity of treating 50 m³ of waste per day, but in 1992-1993, only 3.3% of its capacity was utilized (EPD, 1994b). The average space required for composting one m³ of waste in SLCP is about 0.8 m². The cost of running the SLCP is high. The capital and operation costs add up to about HK$2,700/m³ of livestock waste due to serious under-utilization of the plant. A major reason for the under utilization of the plant is the lack of practical collection system for the livestock waste. The compost is sold at slightly above HK$300/m³. Yet, only 75% of the compost was sold in 1992–1993. The major users of local compost are the landscapers.

There is one private composting plant in Hong Kong. Saw dust, woodchip and chicken manure are turned into compost which is sold at 60% the price of government compost. All the raw compost materials are sourced locally. The open windrow type system is used. The retention time is 75 days to three months and the shrinkage is about 40% by volume. The plant has already undergone contraction due to the unfavourable market sitution. Composting business in Hong Kong, as commented by the representative of the firm. "is hardly viable because the product is low-valued and it involves large overheads" (Agribusiness Ltd., 1994).

Composting is included as a bulk waste reduction option in the proposed "Waste Reduction Plan" (EPD, 1997) The precondition for carrying out large-scale composting is still the availability of markets for the compost products. Marketing of the compost products is also difficult in Hong Kong. There is the need to develop a wider application of compost products in silviculture, surface mine reclamation and landfill final cover. The major constraint is likely to be economic viability since traditional surface filling materials are cheap and abundant in Hong Kong. In the USA most of the materials that are marketed for beneficial reuse are given away, or sold, or sold at a price that does not fully recover transportation costs for the compost facility operator (Gamelsky, 1994). Composts have a limited number of alternative uses, for example, as a sound-proofing material in sound reduction barriers along side motorways (Warren Spring Laboratory, 1992). However, the economic competitiveness of using compost as an alternative means is not good at present.

Climatically, Hong Kong is a suitable place for composting. However, economic viability is the chief factor in handicapping its development. A wastewater collection and treatment system, a large area for windrow composting and the necessary precautionary measures against variable weather conditions, the odour and dust problems will increase the running cost of the plant. If composting is to be viewed as a business or an activity that has to be self-financing then it is unlikely to develop on its own. However, as with other waste management options, composting as a form of tertiary recycling process has the environemental merit of reducing waste volume and producing low environmental impact. These merits must be balanced with the financial aspect.

Lessons for the Developing Countries

In sum, the following lessons can be learnt by the developing nations from the waste management experience of Hong Kong.

Prevention is Better than Cure
Although commonplace, this is not often observed. Remedial and pollution control oriented measures are not enough to deal with the environmental impacts especially depletion of resources resulting from poor waste management. Thus, waste reduction and material recycling and waste utilization (e.g., WTE) are better alternatives to landfilling or incineration with no energy recovery (Chung, 1996). To make waste reduction effective, reforming the waste management system only is simply barking at the wrong tree. The key is to promote sound resource management. Life cycle assessment techniques, product design that enhances reuse and reassembling, incorporation of the producers' responsibility system extensive environmental education and economic incentive in clean production are all effective measures.

Planning of and Equipping the Waste Facilities
Waste facilities must fit into long-term town planning and an adequate buffer zone must be allowed. Since the trend of urban sprawl is generally very difficult to predict with a high degree of accuracy in developing countries, it is important to equip these waste facilities with adequate pollution control devices, e.g. lining, gas extraction system, leachate collection and treatment, flue gas scrubbers, etc. So that even when subsequent urban development around these facilities is unavoidable, there will not be significant health and environmental impact on the people.

Comprehensive Consideration is Necessary

The cheaper option is generally the most appealing option for a developing economy and especially when the plan in question is not of high priority. However, these options are not necessarily the cheapest under comprehensive analysis. This is a major mistake found in the existing waste management plan for Hong Kong. The 1989 plan failed to consider the value of volume reduction in waste, the opportunity cost of land and the potential value of energy from waste. Thus, reaching a conclusion that landfilling is preferred to incineration. This has created an irreversible impact by depleting the precious land resources in Hong Kong.

Similarly, waste recovery has not been fully recognized for its enviornmental value and the waste revovery sector has declined under the government's rigid adherence to the positive non intervention policy. The failure of the decision-maker to consider the environmental gains of resource recovery and landfill space saving are the main reasons for making them believe that the positive non-intervention policy should also apply to the waste recovery sector. In both cases, negligence in the decision-making process is found and irreversible harm or irreparable damage is done to the environment.

CONCLUSION

The development of waste management in Hong Kong has undergone several stages. In the earlier stage, waste facilities were provided simply to satisfy the need for accommodating solid waste and minimal considerations were given for pollution control. In the second stage, improvement in environmental precautionary measures in the waste management facilities was noted but still the waste management authority failed to recognize the more profound need in waste management, i.e., resources management and sustainability. At present, waste authorities are beginning to realize and admit that there are more important goals than simply to reduce pollution from solid waste, but also to conserve resources and energy. Nevertheless, the awakening comes too late. Worse still, the non-discriminate use of positive non-intervention policy has limited the assistance given to the waste recovery sector in Hong Kong. It is hoped that developing countries that are about to improve their waste management system can learn from our mistakes and avoid the detour that we have made.

END NOTES

[1]   The exchange rates between HK$ and other currencies are: US$1=HK$7.8; £1=HK$12.5; ¥100=HK$6.3.

[2] Metal recovery works exceeding 50 kg/hour are controlled by the Air Pollution Control Ordinance. The operator has to licence with the EPD and have to take the best practical measures to ensure no noxious or offensive emission are released to the atmosphere from their smelters.

REFERENCES

Agribusiness Ltd., 1994. Pers. Comm. with Mr R. Au., 15th December, Hong Kong.

Chung, S.S., 1996. Policy & Economic Considerations on Waste Minimisation and Recycling in Hong Kong, Ph.D. Thesis, Hong Kong Polytechnic University, Hong Kong.

Environmental Protection Department, 1986. *Environment Hong Kong 1986.* Hong Kong.

Environmental Protection Department, 1988, Monitoring of MSW, Hong Kong.

Environmental Protection Department, 1993, *Environment Hong Kong 1993: A Review of 1992.* Hong Kong

Environmental Protection Department, 1994, *Enviornment Hong Kong 1994: A Review of 1993.* Hong Kong.

Environmental Protection Department, 1994b, Pers. Comm. with Mr. W.M. Cheung on 27th Oct. Hong Kong.

Environmental Protection Department, 1995, *Environment Hong Kong 1995: A Review of 1994.* Hong Kong.

Environmental Protection Department, 1996a, *Environment Hong Kong 1996: A Review of 1995.* Hong Kong.

Environmental Protection Department, 1996b, *The Environmental Challenges of Hong Kong, Environmental Protection Department 1986-1996.* Hong Kong.

EPD, 1998, *Environment Hong Kong 1998; A review of 1997,* Hong Kong.

EPD, 1999a, *Recovery & Recycling of Municipal Solid Waste in Hong Kong,* Waste Reduction Factsheet No. 1, Sept., Hong Kong.

EPD, 1999b, 1998. Waste update at www.info.gov.hk/epd/Elpub/sw-vep/98update/

Gamelsky, S.M., 1994 "The quest for success in MSW composting", *Solid Waste Technologies,* v. 8, n. 2, pp. 20-27.

Hong Kong 1996, 1996 Hong Kong

Hong Kong Economic Times, 1998, 20 November P.A2.

HK Polytechnic and HK Construction Association Ltd., 1993. *Reduction of Construction Waste: Final Report.* March, Hong Kong.

Kuh, J.R. and Mang. J.L. 1994. "Creating jobs from waste materials" *UNEP Industry & Environment.* April-June. pp.18–21.

Planning, Environment and Lands Branch, 1989, *Waste Disposal Plan for Hong Kong.* December, Hong Kong

Planning, Environment and Lands Branch, 1993, *A Green Challenge for the Community,* Hong Kong.

Planning, Environment and Lands Branch, 1997, *Draft Waste Reduction Plan: A Consultative Paper,* Hong Kong.

Lauritzen E.K. and N.J. Hahn. 1992, Building Waste Generation & Recycling, ISWA Yearbook 1991–1992, Cambridge.

Leung, D.P. 1995. Personal Communication 20th November, Hong Kong.

Stokoe, M.J. 1996. An integrated reduction plan for Hong Kong. *Green Productivity,* 1996: 14–18.

Warren Spring Laboratory, 1992, *A Review of the Environmental Impact of Recycling,* prepared by the Recycling Advisory Unit for the Environment Unit DTI, July, UK.

11

Sustainable Waste Management in Metro Manila, Philippines

Leonarda N. Camacho

Metro Manila Council of Women Balikatan Movement, Inc.
215 Regency Park, 207 Santotan Road, Quezon City 1100
Metro Manila, Philippines

Metro Manila, in the Philippines, is a metropolitan area consisting of 12 cities and five towns. It has a total household population of 1,567,665 million; 1,489 public and private schools, 1,172 subdivisions, 121 public and private markets and about 20 large shopping malls.

The Metropolitan Manila Development Authority is charged with the garbage disposal in the 17 cities and towns. However, presently, the 17 local governments do their own collecting of garbage through hired contractors or through their own trucks.

The collected garbage is dumped in various sites notably in San Mateo, Rizal; Antipolo Rizal; Payatas, Quezon City; Carmona, Cavite. Other towns dump their garbage in rented sites as in Meycauyan, Bulacan; Calmon, Malabon; Sto Nino, Marikina; Pinagbuhatan, Pasig.

The system, which is able to collect only an estimated 50-60% of the entire estimated 5,000 tons daily, is very expensive (P 1, 668, 600.00 billion yearly) and very unsanitary, to say the least.

The dumpsites pollute the air with methane gas and foul odour and the water and land with heavy metals, especially in the area of the water system of Metro Manila in San Mateo and the Antipolo watershed.

The Linis-Ganda (Clean Beautiful) programme is a resource recovery system in Metro Manila which involves the separation of household garbage in the kitchen and the collection of recyclables by a network of eco-aides, for which they receive payment.

The recyclables include: paper and all paper products; all kinds of plastics; whole and broken bottles; car batteries; tin cans; galvanized iron sheets; and all kinds of metals.

The programme is effective in: reducing the pressure on the dumpsites by 50%; conserving natural resources such as trees (as scrap paper is reused by paper mills); oil (as plastics are melted into pellets to make new products), silica (as scrap glass is again made into new glass products) and iron ore (as all metals are converted into billets to make new products); substantive saving to the government in costs of hauling garbage from each household; unquantified savings in fuel of factories of paper, plastics, bottles and cans, as less fuel is consumed when making products out of used scraps rather than virgin materials; savings by the government in importation of raw materials; reducing litter in the streets and canals; reducing air, water and land pollution from open dumping, creating healthier and cleaner surroundings; and protecting the environment and the people.

The programme was initiated in 1983 by the Metro Manila Council of Women Balikatan Movement, Inc, in a small suburban town of Metro Manila called San Juan. By 1992 it was so successful that the Balikatan women expanded the system.

The women organized the junk shop owners willing to join the programme, into cooperatives. Today, there are environment cooperatives in all the 17 cities and towns of Metro Manila. These 17 city/town cooperatives are organized into the Metro Manila Federation of Environment Cooperatives, Inc.—the pioneer for the recycling industry of the Philippines.

The 500 junk shops which comprise the federation in turn employ more than 1,000 eco-aides who collect from each household at least once a week. They also collect once a week from schools, separating the garbage.

Each morning, the junk shop owners give the eco-aides seed money and the routes they are expected to serve. All eco-aides wear green uniforms and use green push carts or green bicycles with side cars. They also carry ID cards. At the end of the day, the eco-aides return with the recyclables and the junk shop owner buys them and gives the eco-aide the difference between the seed money and the cost of the recyclables. The regular income of an eco-aide is $5 to $10, depending on the income level of the community where the eco-aide is collecting. In the case of plush villages where the very wealthy live, an eco-aide might earn as much as $50 a day.

The programme ensures that the junk shop owners have adequate capital to finance the buying from each household by assisting the city-town cooperatives in securing low interest and collateral-free loans from the Department of Trade and Industry and the Land Bank. This is the reason why the junk shop owners were organized into cooperatives because in the Philippines there is the cooperative law which provides that all cooperatives have the privilege of borrowing from government agencies at preferred interest and without collaterals.

The programme provides for free push carts, bicycles with side cars, uniforms and ID cards.

The programme also assists the junk shop owners in contacting prospective buyers of scrap materials such as the 18 paper mills of the country, the five glass factories; the 600 plastic manufacturers; and the five metal plants.

The programme prints the flyers informing the people of the programme and weekly routes and distributes them through village councils and non-government organizations.

The programme has two components: a network that will collect and willing households that will systematically separate garbage.

To monitor all activities, the organizers meets with the officers of the 17 environment cooperatives twice a month. At these regular meetings all problems are addressed and changes made to improve the programme.

Regular seminars are held for the eco-aides and junk shop owners to educate them in the benefits of a clean environment.

Today, the estimated purchase monthly by the eco-aides is 4 million kg. of waste paper, plastics, bottles and cans alone bought from the households for P8 million pesos. Which actually means a saving of collection and transportation of 200 truckloads of garbage on the part of the government, every month. Considering that the Metro Manila local góvernments spend P 1,668,600.00 billion yearly, the savings is staggering.

These figures covers only about 200,000 households who are cooperative enough to separate their garbage.

CHANGES NEEDED LOCALLY

The Metro Manila government should enact a law requiring all households to separate their garbage in the house and allow the eco-aides to buy the recyclables. The Metro Manila government should make all efforts to publicize the system and encourage the citizens to adopt it.

The national government should protect the recycling industry by controlling the importation of finished products such as paper, plastics, glass, cans and billets.

The households, on the other hand, should try to reduce the garbage by buying only what is necessary; carrying their own shopping bags; reusing paper, plastic bags, bottles, etc., and avoiding the collection of packaging that cannot be recycled.

POSTCRIPT

The Metro Manila Linis-Ganda Programme has on its agenda the completion of the following projects by year 2000.

1) The conversion of food and animal wastes into soil conditioner in compost pits or beds to be set up at the San Mateo dumpsites.
2) The conversion of garden wastes and wood debris from construction sites into charcoal, also at the dumpsites.
3) The conversion of animal and food wastes at the market and slaughter houses into methane gas for use at these same markets and slaughter houses.
4) The pelletizing of rubber tyres (an estimated 300,000 unusable tyres annually) for mixing with asphalt.

Unintentionally, the Metro Manila Federation of Environment Multi-Purpose Cooperative Inc. is now emerging as the lead actor in the emerging industry in the Philippines—the Recycling Industry.

Actually, in the existence since 1992 but federated on July 8, 1996, the federation contributes some P100 million to the underground economy annually. The amount represents the cost of the 50 million kg of scrap paper, plastics, glass bottles and steel brought from 200,000 households, schools, and business establishments.

The federation, whose some 500 junk shop owners and more than 1,000 eco-aides are pioneers in economic change, promising an enormous number of new jobs.

The federation harnesses human resources (eco-aides) and in the process reduces consumption of natural resources and energy through the re-use of wastes.

The biggest payoff is in the environment.

Recycling is moving from its supporting role in waste disposal to a preferred method of getting the maximum return from a shrinking supply of virgin resources.

An additional bonus is the savings of some P18 million set aside by the Department of Public Works and Highways to keep waterways, rivers and other tributaries open and clean. Because eventually people would opt to sell their scrap paper and plastic rather than throw them in the canals.

In essence, the recycling industry:
1) reduces air, water and land pollution, especially in the dumpsites;
2) reduces the demand for water used in processing paper, plastics, glass, steel by 50%;
3) saves energy like fuel oil at the factories because recycled wastes melt at lower temperature,
4) saves space in the dumpsites
5) conserves raw materials;
6) saves further destruction of forests, oceans (for oil), mountains (for minerals) and quarries (for silica);
7) keeps the surrounding clean and tidy;
8) gives jobs to many people;

9) saves dollars that will otherwise go to the importation of raw materials; and
10) earns money for households.

Proper garbage disposal is the major goal of the recycling industry. However, the prevention of garbage collection is its larger goal. There is need to reduce garbage. This can be done with concerted efforts. In the recycling industry there is optimism because one cannot afford pessimism.

	City/Municipality	Cost of garbage hauling (annual)	Household polulation
1	Kalookan	130,000,000.00	150,972
2	Las Pinas	24,000,000.00	57,774
3	Makati	120,000,000.00	89,310
4	Malabon	38,000,000.00	58,051
5	Mandaluyong	50,000,000.00	49,065
6	Manila	547,500,000.00	308,874
7	Marikina	48,000,000.00	60,090
8	Muntinlupa	24,000,000.00	53,449
9	Navotas	12,000,000.00	38,864
10	Paranaque	100,000,000.00	61,252
11	Pasay	42,000,000.00	73,642
12	Pasig	48,000,000.00	77,621
13	Pateros	10,000,000.00	9,808
14	Quezon City	400,000,000.00	331,760
15	San Juan	23,160,000.00	24,338
16	Tagig	25,000,000.00	53,153
17	Valenzuela	40,000,000.00	69,642
	Total	P1, 681,660,000.00	1,567,665

(Figures from Metro Manila Development Authority).

Cities/Municipalities	Land area (km²)	Population	Pop. density (person/km²)	No. of Barangays	No. of households
Manila	38.28	1,598,918	41,769	974	308,874
Quezon City	166.25	1,666,766	10,026	140	331,760
Kalookan City	55.81	761,011	13,636	188	150,972
Malabon	23.37	278,380	11,912	21	58,051
Navotas	12.60	186,799	41,825	14	38,864
Valenzuela	47.00	340,050	7,235	32	69,642
Pasig City	12.97	397,309	30,633	30	77,621
San Juan	10.38	126,708	12,207	21	24,338
Marikina	38.94	310,010	7,961	14	60,090
Taguig	33.70	266,080	7,896	18	53,153
Pateros	10.40	51,401	4,942	10	9,808
Makati City	29.86	452,734	15,162	32	89,310
Mandaluyong City	25.96	244,538	9,420	27	49,065
Paranaque	38.32	307,717	8,030	16	61,252
Las Pinas	41.54	296,851	7,146	20	57,774
Muntinlupa	46.7	276,972	5,931	9	53,449
Pasay City	13.97	366,623	26,244	200	73,642
Total	646.05	7,928,867	12,273	1,766	1,567,665

(Figures from Metro Manila Development Authority).

12

Collection, Transportation and Disposal of Municipal Solid Wastes in Delhi (India)—A Case Study

D.K. Biswas, S.P. Chakrabarti and A.B. Akolkar
Central Pollution Control Board
(Ministry of Environment & Forests)
Parivesh Bhawan, East Arjun Nagar, Delhi 110 032

INTRODUCTION

Among the major environmental concerns that Indian cities are confronted with, problems relating to the disposal of municipal (liquid and solid wastes) and vehicular pollution are becoming increasingly menacing. The disposal of domestic sewage and solid wastes and their management is the responsibility of the local authority, the Municipal Corporation or the municipalities.

The city of Delhi, the political heart of India, is spread over an area of 1484 sq km and has a population of 10.3 million. During 1961-1996, the capital city witnessed significant growth in terms of urbanization, vehicular population, industrialization and other related activities.

These activities have become matters of serious concern for the local authorities. The disposal of municipal solid wastes has posed a serious problem because of the fact that there is no proper system to meet the specified standards required for ideal management. Due to the unsatisfactory conditions prevailing in the city, the citizens were compelled to seek assistance of the law for expeditious actions. As a result, one of the legal advocates, filed a Public Interest Litigation during 1994.

In this chapter, the study carried out by the Central Pollution Control Board (CPCB) in compliance of the Supreme Court's direction is presented.

PUBLIC INTEREST LITIGATION ON SOLID WASTE MANAGEMENT

An advocate of the Supreme Court, B.L. Wadhera, filed a case in 1984 against the local authorities, the Municipal Corporation of Delhi (MCD) and New Delhi Municipal Council (NDMC) through a writ petition. It was stressed in the petition that the present system adopted by the MCD/NDMC is not satisfactory as it was creating unhygienic conditions over the city. The Apex Court heard the matter and granted sufficient time to the local authorities for improving the existing management practices.

On March 1, 1996, the Supreme Court issued 14 directions, some of the important ones were:

(a) Lift garbage/waste from the collection centres every day and transport it to the designated places for disposal;

(b) Install an incinerator in all the hospitals and nursing homes, with 50 beds and above;

(c) Appoint Municipal Magistrates for the trial of offences as provided in the Municipal Acts;

(d) Educate residents of Delhi through various mass-awareness programmes and make them aware of the provisions of penalties that exist in the Municipal Acts in case of violations;

(e) Direct the Television Network **Doordarshan** to undertake programmes of educating the residents of Delhi regarding their civic duties;

(f) Immediately purchase vehicles to transport garbage on a daily basis;

(g) Identify areas for future sanitary landfill (SLF);

(h) Revive compost plants and explore the possibility for additional compost plants: and

(i) Take post-operative care for SLF sites.

DIRECTIONS TO THE CENTRAL POLLUTION CONTROL BOARD (CPCB)

The Supreme Court, through its order of March 1, 1996, issued a direction to the Central Pollution Control Board (CPCB), a statutory body created under the provisions of the Water (Prevention and Control of Pollution) Act, 1974, which reads as follows:

> We direct to the Central Pollution Control Board (CPCB) and Delhi Pollution Control Committee (DPCC) to regularly send its inspection teams in different areas of Delhi/New Delhi to ascertain that the collection, transportation and disposal of garbage/waste is carried out satisfactorily. The Board and the Committee shall file the reports in this Court by way of an affidavit after every two months for a peroid of two years.

Thus, the CPCB is bound to comply with the directions issued by the Apex Court.

EXECUTION OF DIRECTION—A CASE STUDY

In compliance with the Supreme Court's direction to CPCB, a programme for study in consultation with the local authorities, Municipal Corporation of Delhi (MCD) and New Delhi Municipal Council (NDMC) was scheduled. After interaction, a work plan was prepared to study the mechanism of collection, transportation and disposal of municipal solid wastes in Delhi.

Meanwhile, MCD consulted the National Environmental Engineering Research Institute (NEERI), Nagpur (India) pursuant to a separate order issued by the Supreme Court to assist the authority in improving the existing system relating to the collection, transportation and disposal of solid wastes.

OBSERVATIONS AND FINDINGS

The management of municipal solid waste in Delhi was mainly looked after by two authorities—MCD and NDMC. The MCD managed of municipal solid wastes through its Conservancy, Sanitation and Engineering Department (CSE), whereas the NDMC regulated such activities through its Department of Health. Out of the total area of 1439 sq km, MCD covered 1397 sq km while NDMC covered 42 sq km. A part of the NDMC area was also being looked after by the Delhi Contonment Board.

The MCD divided its area into 12 zones and NDMC into 13 circles.

Composition of Waste

The National Environmental Engineering Research Institute (NEERI) analysed the characteristics of the municipal solid waste in the MCD area during June, 1996 (Table 1).

It was observed that most of the paper, plastics, glass and metals were picked up by waste picker ("ragpickers") and recycled. The chemical analysis of the waste indicated low organic carbon content in the solid waste where the C/N ratio ranged from 1:20 to 1:30 and the calorific value was between 528 and 875 kcal per kg.

Solid Waste Quantity

It has been estimated that about 4,600 metric tons of solid waste was generated everyday. This estimation was based on per capita waste generation and on the number of waste conveyor trucks reaching the disposal site.

Table 1. Characteristics of Municipal Solid Waste in Delhi (percentages).

Physical:			
	Moisture	—	32.51
	Biodegradable	—	38.6
	Paper	—	5.57
	Plastics	—	6.03
	Metals	—	0.23
	Inert	—	34.7
	Glass and crockery	—	0.99
Chemical:			
	Moisture	—	43
	Organic carbon	—	20
	Nitrogen	—	0.85
	Phosphorus	—	0.34
	C/N Ratio	—	24.08
	Calorific value	—	712 kcal/kg

Collection of Solid Wastes

The system for the collection of solid waste adopted by MCD and NDMC was quite primitive. Waste collecting receptacles in use were masonry structures called a dhalao/dustbin, and trolleys. The dhalao is a masonry structure with a roof and of large capacity (52 to 72 cu m). A dustbin is a roofless masonry structure and at some places it is provided with an iron mesh cover. The NDMC used trolleys which rest on wheels and were lifted by refuse collector.

A detailed survey showed that the entire collection system was not satisfactory because of the following reasons:

a) The maintenance of waste collection centres was not satisfactory. Several recaptacles (dhalaos and dustbins) did not have doors and were in dilapidated conditions. Some dustbins did not have covers and as a result they attracted scavengers.

b) There appeared to be no norm on setting up of waste collection centres. The location of waste receptacles was primarily based on the availability of an area rather than convenience of the public.

c) There was no facility for door-to-door collection of waste and therefore it was the responsibility of an individual to deposit the garbage in the waste receptacles.

d) The waste receptacles were frequently encroached by waste pickers and stray animals. Under such conditions, individuals would hesitate to approach the dhalao or dustbin to deposit garbage.

e) Dustbins and dhalaos proved unhygienic primarily because of spill-overs due to the irregular lifting of garbage.

f) No guidelines had been formulated such as minimum distance to be kept while setting up dhalaos/dustbins so as to facilitate in depositing garbage. The trolleys used in the NDMC areas were old and not compatible with the refuse collector. The open trolleys accumulate water during the rainy season.

Transportation

The most common transportation fleet used included trucks, tippers and refuse collectors. The authorities confirmed a shortage in the transportation fleet. The important findings of the study on transportation of garbage were:
a) All waste-collecting centres were not being cleared on a daily basis.
b) By and large, the loading of garbage in the trucks was a manual process and a limited quantity of garbage was being lifted by using front-end-loaders;
c) Most of the garbage was transported in open trucks without proper covers;
d) The collection and transportation systems were not compatible. The masonry structures, dhalaos and dustbins, were not compatible with fornt-end loaders. These loaders damaged the walls of the receptacles.
e) The trolleys in use were old, and the hydraulic system of the refuse collectors also were not well maintained.

Disposal of Solid Waste

At the time of this study there were three landfill sites in Delhi and the wastes were used for landfilling in an area of 130 acres. On an average, 3600 metric tons of waste used for landfilling everyday and therefore approximately 1000 metric tons per day was disposed of in an disorganized manner. This indicates that in Delhi 100% waste is not collected. The observations on the landfill disposal sites were as follows:
a) the sites had not been prepared for sanitary landfill operations.
b) the bottom of the sites were not pitched with an impervious lining and as a result, there were chances of groundwater contamination due to the formation of leachates.
c) Layer upon layer of garbage was piled up only occasionally covered with topsoil.
d) after spreading on the site the waste was not compacted and the required compaction densities were not maintained.
e) no vents had been provided for the release of inflammable gases such as methane and other landfill gases which normally get released during anerobic decomposition.
f) landfill sites were not fenced, and were therefore encroached by waste pickers, and stray animals.

Composting of Municipal Solid Waste

MCD and NDMC had a few small compost plants. Each plant produced 80 tons of compost daily. The efficiency of the plants, however, was not satisfactory.

Other Observations

(a) Besides disposal of municipal solid waste, other wastes, like horticulture waste, construction debris etc., were not properly disposed of. Haphazard disposal of horticulture and construction waste material encouraged people to deposit garbage indiscriminately at such dumps. Due to inadequate waste collection several garbage heaps remained unattended.

A comparative account on garbage collection, transportation and disposal by the two authorities is given in Table 2.

PUBLIC GRIEVANCES

After the first round of inspection, it was observed that some of the locations in the city could not be covered, and therefore the collected data seemed incomplete. It was also felt that the managment of municipal solid waste is a public subject and unless public opinion was analysed, the study was remain incomplete. With this background, CPCB attempted to seek public participation in the study. By advertising in 17 local Newspapers, the public was requested to report any waste receptacles in any part of city which existed unhygienic condition, or to give any specific observation, to CPCB.

In response to the advertisement, during May-August, 1996, more than 1000 responses were received from all over Delhi. The aspects highlighted by the public were analysed and these views are presented in Table 3.

INTERDEPARTMENTAL COORDINATION

In addition to the MCD and NDMC, other goverment departments are also supposed to be involved directly or indirectly in the management of municipal solid wastes. The agencies involved and actions sugggested on their part are indicated in Table 4.

DECISION OF THE SUPREME COURT

Following the directions issued by the Supreme Court, CPCB submitted three reports which were referred in the Court at three hearings.

Table 2. Solid Waste Management System of Delhi.

Particulars	Municipal Corporation of Delhi (MCD)		New Delhi Municipal Council (NDMC)
Area covered (in sq km)	1399.26		42.40
Population served	4,80,000		12,500
Zonal centres/circles	12 zones		13 circles
Total staff	39,773		2,172
Quantity generated (tons per day)	4600		400
Garbage collection centres (numbers)	Existing	Proposed/ UC	
Dhalaos	337	1252	
Dustbins	1284	708	394
Open Sites	176		
Steel Bins	7		
Trolleys			550
Total:	1804	1960	944
Transport:			
-Refuse removal truck (8 cu m capacity)		431	'Shaktiman' —20 truck tippers
-Refuse collectors (Big) (14 cu m capacity)		65	(3 cu m capacity)
-Mini refuse collectors (6 cu m capacity)		70	Tata truck tippers —32
-Dumper placers (with 5 cu m capacity bins)		59	(3 cu m capacity)
-Mechanical road sweepers		05	Refuse collectors —12
-Front-end-loaders		95	DCM Toyota—08

Table 3. Analyis of public views.

S. No	Subjects of public opinion
1.	Haphazard disposal of garbage
2.	Lifting of garbage irregular
3.	Inadequate dustbins or waste receptacles
4.	Garbage dumping in *nallahs*/open drains by residents, and officials responsible for cleaning
5.	Garbage problem and sewage disposal including open defecation
6.	Dumping of construction debris and garbage in backlanes and on roads
7.	Removal of dead animals, nuisance caused by buffalo owners for dairy and dangers posed by stray animals
8.	Shifting of waste receptacles due to poor maintenance and unhygienic conditions
9.	Dumping of garbage in open vacant plots
10.	Garbage being burnt, trucks carrying garbage plying without cover
11.	Lifting of de-sludged material which is left after cleaning of drains, small drains and water/wastewater stagnation
12.	Suggestions on various aspects of garbage management

Table 4. Recommendations to Concerned Agencies for Solid Waste Management

Concentrated authorities	Suggestions/Recommendations
Delhi Development Authority (DDA)	(1) For proposed residential housing complexes, shopping complexes, the Delhi Development Authority (DDA) should provide proper dustbins and make arrangements to clean them till these are handed over to the concerned authority. (2) DDA Should remove leftover construction debris and this should be ensured on priority. Similarly, arrangements for disposal of horticultural waste should be dealt with. (3) Open plot/area earmarked by DDA for defined activity, such as parks, shopping complexes etc. should not be kept vacant for a long time. Such an act leads to an unhygienic state due to non-disposal of garbage. (4) For the existing DDA colonies and shopping complexes where DDA has failed to provide dustbins and has handed over the area to the MCD, the MCD should construct the dustbin and receive its cost from DDA.
Mahanagar Telephone Nigam Ltd (MTNL)	Any digging work undertaken should ensure that no malba is left over after completion of the work.
Delhi Electric Supply Undertaking (DESU)	Any work which generates debris, or other material should be cleared it.
Irrigation and Flood Control Department of the Delhi Government	After the de-silting and de-sludging operation of storm water drains, filthy material should be lifted and disposed of properly without causing any environmental problem. Wherever possible, drains should be covered and such action when contemplated should cover other factors so as not to affect other related activities.
Central Public Works Department (CPWD) and Public Works Department (PWD)	Public Works Department should ensure that residential areas under their administrative control do not cause problems due to generation of garbage. Roads belonging to these authorities should be well maintained for collection, transportation of disposal of solid wastes.
Railway authorities	Railway authorities should make proper arrangements to maintain railway tracks and adjoining areas under its control for preventing disposal of garbage. Railway authorities should make provisions to regulate disposal of unwanted material on the sides of rail track.

The important proceedings of three hearings are summarized here:

May 7, 1996. "The Court *prima facie* was satisfied that directions issued to the Commissioner, MCD and other officers have not been complied with. The Court issued notices to the Commissioner and other senior officers to show cause why they should not be held guilty of the contempt of the court and be punished suitably".

July 24, 1996. The Commissioner MCD informed that "because of financial constraints, work relating to the management of the municipal solid waste was not satisfactory. The Court after the hearing was satisfied that the National Capital Territory, Delhi Administration is not fully defying the direction of the Court and has not rendered any assistance to the MCD to enable it to comply with the directions of this Court. The Court, therefore, was constrained to issue notices to the Finance Secretary, Delhi Administration and Chairman State Finance Commission to show cause why contempt proceeding be not initiated against them. Since the Court had earlier issued contempt notices to the Chief Secretary (Head of Administration), the Court further directed the Chief Secretary to be present at the court on the next date of hearing".

3. August 27, 1996. The Court observed that "the unhygienic conditions which existed prior to the judgement were continuing though there were marginal improvements. The Court gave time up to October 31, 1996 to substantiate compliance with the orders issued by the Court. The case was listed for November 5, 1996."

PROVISIONS IN THE DELHI MUNICIPAL ACTS

Provisions regarding management of municipal solid wastes were spelt out in the Delhi Municipal Act, 1957 and it was made the obligatory duty of MCD to execute them. Some of the relevant portions of the Act are quoted here:

SECTIONS (350) TO (358)

- Daily cleansing of streets and removal of rubbish and filth is the responsibility of the Corporation;
- Rubbish etc., shall be the property of the Corporation;
- All matters deposited in public receptacles, depot and places provided by the Corporation shall be the property of the Corporation;
- Appointment of receptacles, depot and places for rubbish is to be provided by the Corporation;
- The Commissioner shall provide or appoint in proper and convenient locations, public receptacles, depot or places for temporary deposit of rubbish filth and other polluted and obnoxious matter. This includes provision of dustbins for the temporary deposit of rubbish, provision of vehicles or other suitable means for the removal of rubbish and provision of covered vehicles or vessels for the removal of filth;
- It is a binding on the Commissioner to make adequate provisions for preventing receptacles, depot places, dustbins, vehicles etc. from becoming a source of nuisance;

- Duty of owners and occupiers is to collect and deposit rubbish etc.;
- The Act provides a binding on the owners and occupiers to collect the solid waste generated by them and deposit them in public receptacles at such time as provided by the municipal corporation; and
- It shall be lawful for the Commissioner to take or cause to be taken measures for daily collection, removal and disposal of all filth and keep the city hygienic.

STATUS OF MUNICIPAL SOLID WASTE MANAGEMENT IN THE COUNTRY

A survey was carried out by the CPCB on collection, transportation and disposal of municipal solid wastes in Indian cities. In the first phase of the study, 299 (including 23 metro cities) Class-I cities with a population 100,000 and above were covered. The status relating to collection, transportation and disposal of municipal solid wastes was as follows:

- Total population of 299 Class-I cities (including 23 metro cities) was 139,966,369. Average population density for Class-I cities worked out to 6430 person per sq km.
- The municipal solid waste genetated in 299 Class-I cities was 48,134 metric tons per day. Bombay city had the maximum population and it generated the maximum quantity of solid waste approximating 5,355 metric tons per day.
- The per capita generation of solid waste ranged from 0.1 kg per day to 0.929 kg per day. The average per capita generation of solid waste for Class-I cities thus worked out to be 0.376 kg per person per day.
- Manual collection of solid waste extended up to the 51% are compared to transportation through trucks, which was 48%
- Compostable matter comprised the major component of the municipal solid waste, ranging from 24% to 58% with an average of 35% by weight.
- The predominant method of disposal of solid waste was through landfilling (94%) and the current operations were very crude with only 5% of waste getting processed for composting.
- The collection, transportation and disposal of garbage in bulk quantities was through manual and unhygienic methods.
- The waste dumping sites posed serious hazards to the people living in the neighbouring areas. The groundwater in such areas was liable to be contaminated due to the percolation of leachates.
- Intermixing of faecal sweepings and biomedical wastes with municipal solid wastes remained a continued threat.
- The citywise solid waste generation and mode of disposal is given in Table 5.

Table 5. Solid waste generation and mode of disposal in metro cities in 1993.

S. No.	Metro city	Solid waste collection (t/d)	Treatment (t/d)	Mode of disposal (%)		
				Dumping	composting	others
1.	Ahmedabad	1683	84	95	5	–
2.	Bangalore	2000	200	90	10	–
3.	Bhopal	546	100	20	80	–
4.	Bombay	5355	500	80	20	–
5.	Calcutta	3692	Nil	100	–	–
6.	Coimbatore	350	Nil	100	–	–
7.	Delhi	4000	400	90	10	–
8.	Hyderabad	1566	100	92	8	–
9.	Indore	350	50	85	15	–
10.	Jaipur	580	Nil	100	–	–
11.	Kanpur	1200	300	75	25	–
12.	Kochi	347	Nil	100	–	–
13.	Lucknow	1010	Nil	100	–	–
14.	Ludhiana	400	Nil	100	–	–
15.	Madras	3124	Nil	100	–	–
16.	Madurai	370	Nil	100	–	–
17.	Nagpur	443	Nil	100	–	–
18.	Patna	330	Nil	100	–	–
19.	Pune	700	50	93	7	–
20.	Surat	900	225	75	25	–
21.	Vadodara	400	80	95	5	–
22.	Varanasi	412	Nil	100	–	–
23.	Visakhapatnam	300	Nil	75	25	–

CONCLUDING REMARKS

1. Municipal byelaws provide for safe and hygienic collection, transportation and disposal of solid waste/garbage and the responsibilities are bestowed on the municipalities/municipal corporations for this task. However, implementation of the provisions is increasingly difficult because of the high influx of population into urban areas, the living habits of the people, inadequate infrastructure of the municipal bodies and non-adoption of state-of-the-art technologies for the management of the wastes. Insolvency of these bodies due to the minimal tax structure rendered the system ever-dependent on government subsidies with no initiative to aim for self-sufficiency.

2. Garbage collection and transporation systems, the key factors in the managment, remain greatly neglected. These areas require attention to

achieve satisfactory hygienic management. There is a need to introduce modern methods of collection for which financial assistance of the national and international agencies is required. The transportation system needs to suit the local traffic system taking into account congestion and narrow lanes. Foreign investment on designing better facilities for the transportation in the country is required.

3. In recent years, many developed countries have demonstrated different technologies for solid waste disposal. Private indigenous agencies have also taken initiatives, which are being encouraged. It is necessary to adopt suitable technologies for Indian conditions. Overseas proposals should provide such a system which can be operated by the local municipal agency or through a private contractual system.

13

Public and Private Partnership in the Urban Infrastructures

Pradeep Kumar Monga

Commissioner, Municipal Corporation of Shimla
Himachal Pradesh, India

INTRODUCTION

The solid waste arising from human activities is one of the major environmental problems in urban areas causing extensive pollution and posing a threat to human health. In recent times, the problem of solid waste disposal—both domestic and industrial—has become very acute in almost every city in India because of the lack of resources, unplanned growth and limited disposal facilities. As a result, solid wastes are being dumped in a haphazard manner in various parts of the cities causing severe environmental problems. Urban solid waste management has, therefore, become a nation-wide concern in India.

Shimla, the capital city of Himachal Pradesh, has been experiencing environmental degradation on many accounts. Some of the factors[1] which have contributed towards degradation of Shimla's environment are overpopulation, acute shortage of water and civic amenities, overcrowded roads, heavy traffic and resultant pollution, excessive garbage, unplanned growth and illegal constructions, encroachments on forest lands, and sanitation and sewerage problems which are threatening the very beauty and life of this place. The solid waste arising from human activities in Shimla has become one of the major environmental pollutants causing extensive damage to local ecology, and threat to human health. Urban waste management has, thus, assumed priority in the overall developmental plan of Shimla. There is a general consensus that unless and until immediate steps are taken for the proper management of urban solid waste in Shimla, the future scenario of the city environment is very bleak.

The experience of various cities worldwide indicates that solid waste is not a "waste" in the real sense if appropriate use can be found for it. Apart from the landfill, small quantities of solid waste could be used for composting, aerobic digestion, fuel pelletization and recycling. Proper management of urban solid waste can not only help in the reduction of the quantum of waste to be disposed, but can also generate power, produce organic manure and decrease uncontrolled consumption of resources, thereby generating overall improvement in the city environment and quality of life.

This paper covers various common methods of the disposal of urban solid waste in Shimla, and gives a brief account of the advantages and disadvantages of these methods, along with the genesis of the waste disposal problem. It also highlights the present status of urban waste disposal, and innovative management approaches adopted to regulate and manage the growing urban solid waste in Shimla.

STATUS OF URBAN SOLID WASTE MANAGEMENT IN SHIMLA CITY

Geographical Features

The Municipal town of Shimla, the headquarters of the district and summer capital of India during the British regime, is situated on the range of hills described as the last transverse spur, south of Sutlej of the Central Himalaya. Its mean elevation is 7084 feet above the mean sea level, situated at latitude 31°6′ N and longitude 77°11′. The centrally located ridge runs east to west in a crescent shape with its concave side pointing southward. The extreme ends of the town lie at a distance of six miles (km) from one another. Eastward, the ridge culminates in the peak of Jakhoo, more than 8000 ft. in height and 1000 ft. above the average elevation of the station. It used to be covered with the trees of Pine, Oak and Rhododendron. The five miles long cart road runs around its base. The prospect hills of inferior elevation to Jakhoo close the western extremity of the crescent.

The eastern portion of the town is known as Chhotta Shimla while the extreme western side is called Boileaugunj. An outlying northern spur running at right angles to the main ridge is called Elysium. Three and a half miles from the western end of the station are the outlying hills of Jutogh. The scenery of the Shimla hills is exquisite. The valleys on either side are deep and clothed with thick forests. To the south, the Kasauli and Sabathu hills are distinctly visible. The small mountains can be seen further to the left. To the north and east, a network of mountain chain rises over the valleys of Kullu and Spiti in the north, and the central range of the eastern Himalaya stretches in the east and south east.

Geologically, the lower stratum of Shimla hills consists of limestone, grit conglomerate and slate. The upper stratum is metamorphic with highly foliated schistose stone. In parts, mica schistose predominates and on the

Boileagunj side, they are siliceous. A recent survey revealed that the most of land surface along the hill slopes is covered with the unconsolidated permeable overburden soil. The average slope along the northern and western side of the ridge is about 36°. The soil cover along this slope shows signs of slow creep. This sliding tendency of the soil blanket is augmented under an oversaturated condition, especially during the heavy rains in the monsoon months. The free movement of groundwater is obstructed by clay matrix present in the soil, which results in the building of water pressure followed by soil erosion due to its downward movement. The northern face of the slope receives very little sunshine, and is therefore less green.

Demographic Indicators

Shimla is a typical example of colonial growth before 1947, and haphazard and unplanned urban growth after independence[2] (Maitra, 1994). The town was developed by the British so as to recreate their cultural landscape upon Indian lands. Since the beginning, the town has seen many social, political and cultural changes. The present-day Shimla gives the impression of a living organism with a synthesis of the past and present. The demographic features show that population distribution pattern is not uniform, and manifests a tendency of being concentrated on the top of several spurs. There are many areas within the limits of the town which are absolutely devoid of human habitation. These are generally made up of valleys and slopes with high drainage density.

The population density is maximum in the southwest portion of central Shimla i.e. from the Railway station to Combermere stream (Fig. 1). This area has locational advantage of plenty of sunshine, gentle slopes, low drainage density and high degree of accessibility. There are multiple clusters of population around the main Shimla ridge which also acts as a dividing line between the basin of two major rivers—Sutlej and Yamuna. On the northern side, clusters of settlements cover the areas of Longwood, Kaithu, Shankli and Summer hills. The other population clusters are on the southeastern end of the town, i.e., Chhotta Shimla and Kasumpati. On the western side, Tutikandi and Khalini areas are located on the midsouthern periphery of Shimla. Besides, the municipal areas, the development activities are on a sharp increase just outside the limits of municipal area, i.e., in Vikas Nagar on the southeastern side, and in Kachi Ghati, Cemetery, Kasumpati areas on the northern side.

The city of Shimla which was originally designed for a population of 25,000, within the area of 19.55 sq. km. has grown to support 82,054 inhabitants in 1991, and about 1,25,000 people in the year 1995-96 (Figs. 2-4) (Table 1).

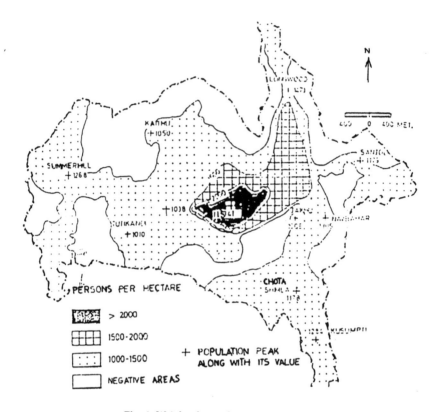

Fig. 1. Weighted population density in Shimla

Table 1. Democraphic features of Shimla municipal corporation.

Census	Population			Pop. density	Occupied residential	Households inc. houseless
	Male	Female	Total		house	household
1981	42254	28330	70584	3611	17500	17743
1991	46854	35220	82074	4197	29826	21023

Source: Census of India Report, 1991.

The city of Shimla had a steady growth of population, construction activities and the pressure of other developmental works since independence. However, the basic civic amenities like water supply, roads/streets, sewerage system, although augmented, are hopelessly lagging behind, and are unable to cope with the needs of the present-day population of Shimla. The population density of Shimla has reached to about 6,250 persons per sq. km. The city being an important hill resort, during the

SHIMLA

POPULATION PROJECTIONS
(Permanent Population)

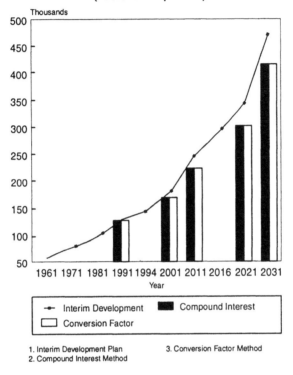

1. Interim Development Plan   3. Conversion Factor Method
2. Compound Interest Method

Fig. 2.

summer months including Dussehra, Diwali and Christmas holidays, a large influx of population takes place putting extreme pressure on the civic amenities. Thus, the average population of Shimla town ranges between 1.5 and 2.5 lakhs at any point of time with peak rush during the summer months.

Quantity and Quality of Garbage

No proper study has been conducted so far in Shimla to assess the quality and quantity of garbage produced every day. However, a preliminary survey revealed that the estimated quantity of generation of garbage is about 0.4 kg per capita per day in Shimla which comes to around 50 to 60 tons of garbage per day for the whole city (Table 2).

Shimla is divided into six sanitary wards and a market area. Each sanitary ward is further divided into beats supervised by a sanitary

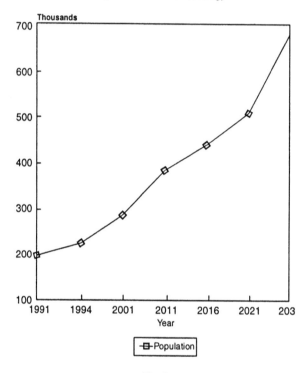

SHIMLA
PROJECTED POPULATION
(Permanent and Floating)

Fig. 3.

Table 2. Garbage production per capita per household in few localities of Jakhoo in Shimla.

Name of localities	No. of households	No. of persons	Total garbage	Garbage/ capita	Garbage/ household
Oakwood	49	221	117.5 kg.	.58 kg.	2.3 kg.
Daizy Bank	38	189	62 kg.	.32 kg.	1.6 kg.
Total	87	410	179.5 Kg.	.43 kg.	2.06 kg.

Source: Report of H.P. Pollution Control Board, 1994-95

supervisor. A total number of 550 sanitary workers are engaged in the job of scavenging, collection and disposal of garbage, upkeep of public toilets, cleaning of drains, and disposal of night soil. The manpower per thousand population works out to be three to four hours and accordingly, eight sanitary workers cater to an area of 1100 sq m³.

SHIMLA

PROJECTIONS FOR TOURIST POPULATION
AS PER INTERIM DEVELOPMENT PLAN

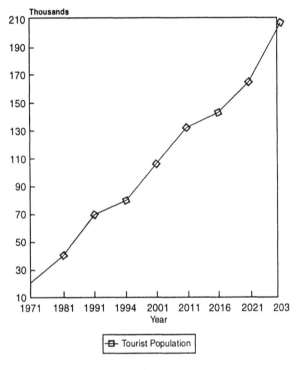

Fig. 4.

Problem Areas

The Shimla Municipal Corporation has provided 500 dustbins and 51 dumpers in the city. The bins are made of cement concrete or fabricated metals boxes, which are not covered. Dumpers which are mechanically lifted, are gererally placed in areas where the garbage load is maximum. The capacity of one dumper is 4.5 cu m., while the capacity of metal and concrete bins is one cubic metre. However, the bins and dumpers are still inadequate to meet the increasing demand. Most of the residents as well as the tourists are in the habit of indiscriminately throwing garbage packed in polythene bags on hill slopes and in the drains, leading to frequent blockade of drains besides giving an ugly look to the hillsides. The increasing pressure of non-biodegradable waste polythene bags and trash is a continuous source of environmental degradation and health hazards in Shimla. Of the 50% of total garbage generated only 50% goes into the dustbins, and rest in thrown towards the hillside. The main reasons for the indiscriminate disposal of garbage on the hillocks and drains can be partly attributed to the lack of public awareness, non-availability of dustbins, and irregular clearance of dustbins. The urban waste and garbage is usually disposed of by throwing it on hill sides, using landfills, and by open burning.

The garbage from the dustbins to the final disposal site is carried by the Municipal Corporation vehicles. The frequency of collection at locations which have the maximum generation of garbage, is on a daily basis, while at other locations, it is on alternate days or twice a week. Out of 500 dustbins and 51 dumpers scattered all over the city, only 50% on an average are emptied every day. Further, only 70% of the entire city area can be covered by these vehicles as there is no approach road in some parts of Shimla. There are several areas outside the Municipal Corporation limits having prolific growth but no facility for either proper disposal or collection of garbage. In the past, no scientific method of final disposal of garbage was adopted, and the garbage was being disposed of indiscriminately on the Shimla Bypass Road. The final disposal of garbage is known to be a major contaminant to the water drained into the various natural springs. There are about 145 drains which get drained into the two major river basins. Most of the solid garbage which is thrown on the slopes and gorges finds its way into these drains, thereby contaminating nearby water sources.

In brief, some of main problem areas identified in urban waste management in Shimla are: (i) unplanned growth and over-population, (ii) inefficient collection, (iii) lack of segregation of garbage into biodegradable and non-biodegradable waste, (iv) improper and uneconomical transportation of collected garbage, (v) unscientific disposal of garbage, (vi) lack of people's participation, and (vii) lack of resources.

INNOVATIVE PRACTICES IN URBAN WASTE MANAGEMENT

Keeping in view the resource constraints, inadequate manpower, and lack of general awareness, the Municipal Corporation of Shimla has embarked upon various innovative management practices to tackle the problem of Urban Waste Management on a priority basis. These practices were selected after the identification of problem areas, and an in-depth analysis of possible viable options available before the Corporation, after undertaking detailed discussions with all the concerned people/agencies involved in the city waste management process, i.e., the State Government, Elected Representatives, General Public, Scientific and Academic Community, Public and Private Organizations, NGOs, and employees of the Corporation.

Ensuring People's Participation

People's participation and active cooperation has been identified as key elements in Waste Management Strategy being implemented in Shimla. There are many hurdles in ensuring people's participation in collection, transportation and disposal of waste and garbage in Shimla. The lack of awareness, substantial floating population, and cultural and economic background have been identified as some of the main factors contributing to the general apathy to garbage and waste disposal problem in Shimla. The operational gaps have been identified mainly in the area of collection and segreation of garbage in Shimla. Based on this premise the Municipal Corporation of Shimla has evolved and implemented a plan of action last year, i.e., in 1995-96 for involving the general public in waste management programme in Shimla.

Seven steps were involved in the preparation and implementation of the plan of action for effective waste managment in Shimla.

Step 1: Identification of the area of operation. The whole city of Shimla has been divided into 21 operational wards. In each operational ward, area-specific problems in the collection of garbage were identified and target groups were selected.

Step 2: Formation of 300 groups (eco-clubs) by choosing people and organizations having interest and experience in garbage and waste disposal.

Step 3: Brainstorming and extensive discussions to prepare concrete plans for action from abstract generalizations on status of issues, and a general approach to tackling issues like collection and transportation.

Step 4: Entrusting duties and identification of area of operation to the different groups of people i.e., eco-clubs, government/non-government organizations, and experts in the field to work on each issue in all its dimensions. These groups address themselves to these issues by collecting household information, probing the problem areas, elucidating different approaches towards a viable solution, and sharing responsibilities in a coordinated manner.

Step 5: In-depth analysis of past practices and approaches to city waste management, and reasons for their success or failure in the field conditions.

Step 6: Training of target groups, preparation and distribution of literature on various aspects of garbage collection and disposal, and the capacity building of Corporation employees.

Step 7: Formulation and implementation of a new strategy based on past experiences and information collected. This strategy calls for a renewed partnership between the Corporation and general public for sharing responsibilities and resources for effective waste management in Shimla.

Under the proposed strategy, each eco-club becomes the focus of all eco-friendly activities taking place in that particular area. Although garbage collection and dispoal remain the main areas of operation for all eco-clubs, they also play the role of nodal agency in the implementation of various environmental programmes which include tree plantations, community hygiene and civic sense, and protection of greenery and vegetation. Each eco-club (with about 50 households) has to contribute Rs. 500 per month which comes to Rs. 10 per household per month, and the Municipal Corporation provides the equivalent grant of Rs. 500 per month to each eco-club. Thus, each eco-club has Rs. 1000 per month as a grant for employing one part time person who will primarily be responsible for the collection of separated garbage from each household every day, and putting it in the bins/dumpers specifically provided by the Corporation. The role of these part-time employees is supplementary to the work undertaken by the Corporation, and fills the gap between household and dustbins. These eco-clubs are also expected to play a key role in disseminating the information on hazards of using non-biodegradable materials, the importance of recycling and reuse, and the need to plant more trees in and around Shimla.

Public-NGOs Partnership

Involvement of non-governmental organizations (NGOs) has been identified as a crucial factor in the successful implementation of any public programme for better city waste management in Shimla. Since NGOs have the inherent advantage of having close proximity to the gerneral public, they can effectively act as a bridge between the governmental agencies and the general public. The task of organizing eco-clubs, training target groups, preparing literature for increasing awareness on environmental issues and organization of camps, cleaning campaigns and events like 'Run for Nature' involving school and college students and voluntary groups has been entrusted to an NGO in Shimla, the Indian Environmental Society. Linkages have also been made with other NGOs like Gyan Vigyan Samiti, Shimla, Tata Energy Research Institute, New Delhi, The Exnora, Madras, and Centre for Environment Education, Bangalore for exchange of information on urban waste management.

Public-Private Partnership

Public-Private Partnership has assumed special importance in recent times in the field of urban waste management, especially keeping in view the resource constraints and inadequate manpower availabe with the local bodies. Although the Municipal Corporation of Shimla has imposed a sanitation tax in Shimla, the tax collection as compared to the total expenditure on sanitation is almost negligible. This has resulted in a huge gap between the resources and the work to be carried by the Corporation. Lack of adequate resources along with limited manpower has resulted in a sort of a crisis as far as garbage disposal is concerned in Shimla. This has eventually led to privatization of some of the aspects of garbage collection, disposal and general hygiene in Shimla.

Sulabh International Ltd. a non-profit organization has joined hands with the Municipal Corporation of Shimla to manage some of the public convenience facilities and garbage collection in Shimla. Sulabh International Ltd. charges reasonable prices for use of public convenience facilities from the general public, and in return provides a cleaner environment. In addition, garbage collection on contract basis is also being explored by this organization. The Excel Industries, a pioneer private company involved in the city waste treatment and its bioconversion into organic fertilizer is collaborating with the Shimla Municipal Corporation to convert city garbage into useful manure. At the grass-root level, rag pickers have joined hands with the Corporation in the collection and disposal of polythene bags and recycling of other trash.

Scientific Managment and Disposal of Waste

The scientific management and disposal of urban waste has recently become very important in view of various new clean technologies now available for the recycling and reuse of urban waste. Incineration, pyrolysis, conversion to biogas, refuse-derived fuel, conversion to single-cell protein, fuel pelletization, composting and vermicomposting are some of the known scientific methods and technologies relevant for the efficient use and eco-friendly disposal of urban waste.

The Municipal Corporation of Shimla has embarked upon four projects for waste management in Shimla namely: (a) Bioconversion of city waste in organic fertilizer, (b) Energy from urban waste through methanogenation, (c) Incineration of hospital waste and hazardous material, and (d) Recycling of polythene, paper and other useful refuse. In the case of bioconversion of waste into organic manure, the Corporation has entered into an agreement with the Excel India Company which has perfected the technique of bioconversion of biodegradable waste into useful manure by inoculating garbage with artificially raised bacterial culture. This method reduces the normal period of bioconversion of garbage into manure from six to eight months to 45 days. The Municipal Corporation of Shimla is also in the process of selecting a suitable local partner—a private party/entrepreneur/ NGO for taking up the operation of this project on commercial lines. In the case of energy from waste, the Corporation is setting up a methenogenation pilot plant with financial and technical assistance from the Ministry of Non-conventional Sources of Energy, Government of India. To tackle the problem of hospital waste and hazardous material, the Municipal Corporation is putting up two incinerators of 5 tons capacity per day. To recycle and reuse polythene bags, used paper, and cloth rags, the Municipal Corporation is collaborating with the Jan Sewa Ashram, an NGO in Himachal Pradesh which is involved in environmental activities including the production of recycled paper and other useful products from city waste.

Training and Capacity Building

Training and capacity building of the elected representatives, administrators, planners, experts and employees have been identified as key factors for the ultimate success of any urban waste management strategy which calls for partnerships among local authorities, the governments, NGOs, and the private sector. Realizing its importance, the Municipal Corporation of Shimla has started a comprehensive training programme for all the concerned people involved in the urban waste management. Human resource development has been recognized as the key element in the capacity building, and a special programme to sensitize the sanitation workers, eco-clubs, and NGOs in modern scientific methods and innovative practices in urban waste management has been started by the Municipal Corporation of Shimla. Although it is at infancy stage, it has already started yielding results.

CONCLUSION

Urban Waste Management has emerged as one of the important areas of overall urban policy planning and development which requires innovative practices and approaches suited to the local conditions in order to overcome the financial, administrative and operational constraints. An essential component of any urban waste management strategy is ensuring that the local authorities develop the necessary professional skills and capacity. This also calls for a close cooperation between local authorities, the governments, NGOs and private sectors. Keeping this in view, the Municipal Corporation of Shimla has embarked upon a number of innovative practices to tackle the issue of urban waste and garbage disposal effectively, and to provide a better, cleaner and greener environment to its citizens. The main recommendations on the basis of field experience call for active participation of people at every stage of waste management, urgent need to commercialize waste management technologies, close public-private-NGOs partnership, and emphasis on training and capacity building of all key actors involved in urban waste management.

REFERENCES

1. These factors were identified in the National Workshop on Sustainable Development of Urban Habitat-Shimla 2020 held at Shimla, Himachal Pradesh form 1-3 November, 1995.
2. Maitra, A.K. 1994. Environmental impact of urban development in hill towns: A case study of Shimla. Project paper in Spatio-Economic Development Record, New Delhi
3. This is based on the presumption that the total area to be cleared comprises 130 kms. long and 4 mts. wide roads, and around 13 kms. of long drains and other small lanes in congested localities.

14

Urban Waste Management in Hilly Towns

Virinder Sharma

Principal Scientific Officer, State Council for Science, Technology & Environment,
34 SDA Complex, Shimla, H.P., India

INTRODUCTION

The concept of Solid Waste Management is based on a combination of source reduction, recycling, waste combustion and disposal. This encompasses the purposeful systematic control of the functional elements of generation, waste handling, separation and processing at the source, collection, separation, processing and transformation of solid waste, transfer and transport and disposal associated with the management of the solid wastes from the point of generation to final disposal

URBAN WASTE MANAGEMENT IN HILLY TOWNS

Urban Solid Waste Management in the hilly regions of the Country and specifically in Himachal Pradesh requires an approach which takes into account the unique and diverse social, economic and environmental characteristics of the mountain regions. The fast expanding hill towns are now facing the problems of Management of the Solid Waste. The report of the High Power Committee of the Planning Commission on Solid Waste Management has been able to highlight the problems of the solid waste management in terms of collection, transportation and disposal of solid waste. This report has however not taken into account the special problems being faced by the Urban Hill Towns, in terms of the collection and disposal of solid waste. It is seen that the growing menace of solid waste including both Biodegradable and non-biodegradable was not just despoiling the environment, but also affecting people's health. Thus to reduce the pressure on natural resources by adopting recycling methods is the major GOI as well as State priority.

SECTORAL CHALLENGES

Rising population rapid urbanisation and increasing industrial activities are contributing to huge accumulation of wastes in different forms. The waste which is generated from these activities is not managed properly. The waste can be changed to form of energy by putting appropriate technologies in the waste management. As we know that conventional sources of energy are depleting very fast, so at this time the non-conventional sources of energy can solve the problem of natural resource depletion.

In the State of H.P., one of the major problems i.e. emerging due to the present development strategies is the increasing generation and scientific disposal of solid waste in the major towns and villages, With the increase in per capita generation of solid waste and also a corresponding change in the type of garbage [from biodegradable to non-biodegradable and infections waste], Solid Waste Management is emerging as the focal area under environmental conservation and also for improving the quality of life in tribal, temperate and subtropical regions of the state. In Himachal Pradesh, major component of waste is plastic bags which are responsible for environment degradation. Thus one priority of State is to find out methodologies for proper management these plastic bags and other wastes.

MOUNTAIN SPECIFIC PROBLEMS OF SOLID WASTE MANAGEMENT

1. Availability of open slopes where municipal wastes are indiscriminately thrown. These open slopes of provide some aerobic decomposition of these wastes into the soil but are now littered with polythene bags which prevent any further decomposition of the contained organic waste matter. The generally prevalent low temperature averages only serve to aggravate the problem further.
2. The retention of wastes on the slopes causes further problems as these wastes enter into the drainage system and pollute the water sources specifically of the lower lying rural areas.
3. The polythene bags have created problems by choking the sewerage and drainage systems and also by virtue of the fact that polythene is non-biodegradable, thereby remaining on the slopes as such for years. Because of the durability and cost factor, plastic has entered into urban areas as well as rural areas. The increased use of plastic bags has led to mass disposal of garbage in these bags, specially in the semi-urban and urban regions of Himachal Pradesh.
4. The municipal wastes are not easily collected because of the problem of access [transport and roads]. Also the ragpickers and kabaries

[traditional waste picker categories] cannot collect these wastes from steep slopes.

5. Disposal of Solid Waste is presently being done on the slopes as large landfill sites are not available.
6. Recycling of used and thrown coloured polythene bags is not economically viable for the waste pickers in the hills because of the effort it would entail and the cost of transportation.
7. This massive and yet scattered dumping of wastes in plastic bags has led to offensive eyesores, causing increased pollution in the soil as well as the water sources.

It has been estimated that 90% of the problem of waste littering is due to the cheap coloured polythene carry bags. The led the H.P. Govt. to prohibit the sale and littering of the low grade/quality coloured and recycled polythene bags used by vegetable vendors and grocers, which cover 90% of the non-biodegradable plastic material and pose an environmental and health hazard in our daily life.

PROGRAM INITIATIVES IN HIMACHAL

The Council for ST&E has taken many initiatives under Solid Waste Management Programme, for the socioeconomic development of the people and to create healthy environment in their own premises by recycling waste. But the need of the hour is to upgrade the existing technologies for the betterment of solid waste management requiring technical and financial assistance.

OBJFCTIVES

The objectives of the solid waste management programme are as follow:—

1. Waste Survey and Mapping of the towns. This involves participatory survey of garbage collection, transportation and disposal system with the formal and non-formal sectors to develop Waste Management Plan in each locality/ward.
2. Involvement of non-formal sector of Kabaries and Rag pickers in the planning and implementation of SWM.
3. Segregation of biodegradable, non-biodegradable and biomedical waste at source.
4. Provision of paper recycling and plastic bags reusing.
5. Composting of biodegradable components of the waste into organic manure through aerobic composting, microbial conversion and vermiculture through research and development and technology assessment.

6. Suitable disposal of biomedical and clinic waste through incineration or other methods.
7. Information, Education and Communication campaign to create awareness among policy makers, planners, cross sectoral field staff, NGO's and general public.
8. Implementation of the Act through inspections, fines and other methods like seminars/workshops/awareness campaign.
9. Training course for urban local bodies and other technical staff which are working in Waste management programmes.
10. Training and capacity building programme for Rag pickers, scavengers, sweepers, kabaries, NGO's etc.
11. Information Cell for Solid Waste Management programme.

IMPLEMENTATION PROCESS

Himachal Pradesh has taken the lead among all States in prohibition on the sale and littering of coloured recycled polythene bags and also formulation of an Act on Solid Waste Management. The programme started in 1922 by the State Council for Science, Technology and Environment has now been extended to all areas of the State and sensitization of the policy makers, planners and implementing depths have been achieved in the area of Solid Waste Management. This has also resulted in enhancing the capacity building of the formal and non formal sectors, coupled with technical and financial resources being made available by the State Govt. and National Agencies [NORAD Environment Cooperation Project] for improving Solid Waste Management services.

This programme has now emerged as a Success Story in Solid Waste Management all over the country. The Success Story is based on Administrative, Technical and Management Innovation; Govt. -NGO-Local Body Partnership; Legislative Frame Work; Sustainability and Replicability.
The Details are as Follows:

1. INNOVATION

(A) H.P. Non-Biodegradable Garbage (Control) Act 1995 and Rules 1996—Administrative Innovation and Legislative Frame Work

Himachal Pradesh is the first State in the country to have enacted an Act for dealing with solid waste management and the menace of plastic carry bags. The H.P. Non-Biodegradable Garbage [Control], Act was formulated by the Dept. of Science, Technology and Environment, Govt. of H.P. and enacted in 1995, to prevent throwing or depositing non biodegradable garbage in public drains, roads and places open to public view in the State of Himachal Pradesh.

Some of the salient features of the act include-Prohibition to throw non-biodegradable garbage in public drains and sewage, provision for placement of receptacles and places for deposit of non-biodegradable garbage etc. Although, the Act got a lot of publicity initially, the fact is that it remained restricted to littering of plastics. However, It was able to sensitize the enforcement agencies on the issues of solid waste managements. This was an innovative idea of State Government to tackle the nuisance of littering of polythene bags and other non-biodegradable garbage through this Act.

In the year 1998, the Act and Rules were made more stringent and notifications were issued for compounding the offences, and from 1st January, 1999, this Act was made operational throughout the State, including prohibition on the sale of recycled and coloured polythene bags.

Prohibition on Coloured Recycled Polythene Bags
The State Govt. has prohibited the sale of coloured recycled carry bags from 1-1-99 in whole state. Under Section 7 [h] of the Act, prohibition was imposed on the traders retailers and vendors in the State of Himachal Pradesh for using the coloured polythene carry bags manufactured from recycled plastic, for packaging the goods traded/sold by them, with effect from 1st January, 1999.

The Field officers of Urban Local Bodies and Food and Supplies Dept. within their jurisdiction, were authorized to compound any offenses punishable under Act. The shopkeepers were asked to clear the existing stocks of coloured recycled polythene carry bags before the 1st Jan. 1999 and to use the transparent first quality carry bags w.e.f. 1st Jan 1999. For offences in violation of the notification 7[h] of the Act, a minimum amount of Rs. 50/-as compounding was enforced by the authorised officials. An appeal was also made by the Hon'ble Chief Minister to educate the people ill effects of the coloured polythene bags made from recycled plastic waste.

In view of this ban, State Council for ST& E organised different activities. One day Workshop on Solid Waste Management and Ban of recycled coloured Polythene carry bags was organised in Shimla in the month of December in which 100 participants took part. Similar type of One day Workshops were organised in all District Headquarters with H.P. State Pollution Control Board and Dist. Administration Shimla, Bilaspur, Mandi, Kullu, Hamirpur, Kangra, Una, Solan and Nahan. This was followed by General Awareness Campaigns in the State to sensitize the District Administration and the general public about the ban on sale of plastic carry bags. It was decided that the District Administration will monitor and coordinate the activities relating to the implementation of the ban by organising cleaning campaigns, awareness programmes and proper disposal of the existing stocks as well as propagate use of eco friendly products in their districts. Districts also decided to give special incentives to Mahila Mandals, Dwcra Groups to make paper bags, Jholas etc.

Polythene Cleanliness Campaign

The Polythene Cleanliness Campaign started on 5th Dec. 1998 from the SDA Complex area for one month. Aprons marked with the Campaign message and other equipment were given to 20 Kanchan Karmies [ragpickers]. To dissminate the cleanliness drive message among the people, the Kanchan karmies distributed the Appeal Pamphlets in the respective areas of their campaign. The Polythene Cleanliness Campaign was coordinated with Municipal Corporation, Shimla, which also employed the services of 50 sanitation staff. The State Council provided 50 aprons and other equipment to the M.C. staff.

(B) Micro Waste Recycling Units For Schools and NGOS—Technical Innovation

The State Council for Science Technology and Environment had advertised a Solid Waste Management Scheme in H.P., under which Project Proposals were invited from Educational Institutions and NGOs for setting up Micro Waste Recycling Units and Resource Centres for demonstration and awareness of waste recycling. The objective was that recycling of waste paper, waste plastic and other garbage could be initiated by children in their school campus or by NGOs in their area and this activity could be integrated to the SUPW module in the schools. This Resource Centre includes Paper Recycling unit, Micro Polythene Reusing unit and Composting unit in the selected schools where demonstration can be given in converting waste paper and polythene into a useful recycled products through the involvement of children. Total of 12 recycling units were sanctioned for schools in Shimla, Solan, Chamba, Hamirpur, Bilaspur, Kullu, Kinnaur & Lahaul and Spiti districts, out of which five units were set up in Shimla and remaining are being set up in each of the districts.

Teacher Training Wrokshops on Waste Recycling—Management Innovation

The State Council organised two Teacher training workshops on waste recycling in Acukland House School in which 40 participants from different organisations took part in this workshop. One teacher and one helper were nominated by the schools, where recycling units were to be set up, for this three day training on the paper recycling, plastic reusing and composting methods.

The Council received funds under NORAD project for the popularisation of recycled material under which NGOs in districts were assisted in setting up Micro recycling units for Waste paper, plastic and waste composting. This included training of local youth in Waste Recycling modules at these Centres.

Four NGOs were identified for this programme were Samarpan, Chamba; Astha, Hamirpur; Dhauladhar Educational Society, Yol and Sathi, Sirmour.

Meeting With Kabaries and Safai Karamcharies of Shimla

The Non formal sectors of Kabaries and Ragpickers as well as the Sanitation staff [Sweepers] participated in the meetings to formulate an Action Plan based on their feed back for improving Solid Waste Management services and dealing with the problems of polythene bags.

(C) Secretariat Waste Paper Management Scheme—Management Innovation

The State Council for Science, Technology and Environment, initiated a Scheme on Waste Paper Recycling in the H.P. Secretariat. The objective of the scheme is to encourage segregation at source for collection of waste paper and promote the use of recycled paper products. This scheme is being executed through the Jan Sewa Ashram, which is an NGO in the field of Waste Recycling in Himachal Pradesh. This was the first scheme of its kind in the country. Under this programme, 40 Green Bins, one paper shredder and 100 Waste Paper Bins made of Lantana (an exotic weed plant in the state) are being placed in the individual rooms of the Secretariat Office, so that waste paper is directly segregated in the rooms. This will also help the Secretariat Cleaning Staff to collect separated waste paper from individual rooms. The waste paper from the Secretariat Office rooms is segregated from other garbage and put in the Green Dustbins provided in each floor of the Secretariat. The waste paper is being collected and trasnported to Jan Sewa Ashram in Jabli, where this waste paper is recycled into useful paper products like file cover, letter pads etc.

Before this scheme, waste paper which was collected by the sweepers was generally burnt on the slopes nearby the Secretariat. The ash was thrown below the Secretariat Office slopes proving a hazard to trees along the slopes and burning of paper also created smoke and ash hazards/pollution.

(D) Participatory Waste Survey and Assessment—Management Innovation

The state Council conducted Waste Survey and Assessment in five towns of the State viz. Solan, Mandi, Nahan, Chamba and Dharmshala under its Solid Waste Management Programme. The Waste Survey and Assessment was aimed at:

1. Current system of Waste management in terms of types generated, Composition of Waste, means of collection and disposal.
2. Role of the formal sector [Municipal Council Sanitatoin Staff] and the non-formal sector [House hold sweeper and rag pickers].

3. Waste Management plan for door to door collection, segregation of different types of wastes and collection mechanism including disposal and recycling of appropriate category of wastes like paper, polythene and compostable material.

On the basis of the surveys the Solid Waste Management Plan was formulated for Solan town [Bio-conversion Plant set up] and Dharmshala town [Door to Door Waste collection System initiated].

(E) Bio-Conversion Plant [Waste Treatment] at [Solan]—Technical Innovation

Under Solid Waste Management Programme a bio-conversion plant was set up at Solan for garbage processing. MOU has been signed by the state Council, Municipal Council Solan and Jan Seva Ashram [NGO] for setting up this plant. The plant was inaugurated by the Hon'ble Chief Minister of Himachal Pradesh on 28th June, 1998 at Salogra. The Bioconversion plant is the first plant of its kind in Northern India with a capacity of 20 MT of garbage per day. State Council for STE has conducted garbage survey in Solan towns to identify the generation, type and quantum of solid waste in the town which helped Municipal Council and Jan Sewa Ashram in project implementation. The plant is converting Solan town Garbage into compost through Microbial process based on Excel Technology and steps are being taken to popularise the use of this manure among farmers and Govt. Agencies.

2. PARTNERSHIP—GOVERNMENT, NGOS AND LOCAL BODIES

The State Council for Science, Technology and Environment initiated collaboration with the NGO's and local bodies in the Solid Waste Management programmes to formulate the project and assist in implementation, both financially and technically. State Council partnership with NGOs, Government and Local Bodies is a major success in disseminating different SWM technologies up to grassroots level in the state.

3. SUSTAINABILITY AND REPLICABILITY IN OTHER STATES AND HILLY AREAS

The H.P. Act has been further been enacted in the states of Haryana, Punjab, Sikkamand Arunachal Pradesh and workshops have been organised to popularise green purchasing and developing recycling projects by installing Recycling units in these States.

4. IMPACT ON THE PEOPLE—LOCAL COMMUNITY PARTICIPATION

Enforcement of the Act will necessitate educating the people, creating awareness among them and activating the Local bodies which will be called upon to play a leading role. After the notification of the H. P. Non biodegradable Garbage Control Act and Rules in the State the general publicity through the media was that State has banned polythene bags.

From the New Year's day, Himachal Pradesh has become free from Coloured Recycled polythene carry bags. The districts of Kullu, Shimla, Sirmour, Mandi and Hamirpur have taken important strides in the implementation of this Act, while the other districts are gradually becoming effective. Most of the traders, shopkeepers, retailers and fruit and vegetable vendors in the State have switched over to either paper bags, cloth bags or transparent polythene bags.

The people in general strongly feet that ban is the only solution to deal with this problem and they have supported the ban to achieve a clean and green Himachal. Thus the pioneering move of the Himachal Pradesh Govt. banning plastic bags to keep the cities and towns free from the menace of polythene bags has attracted attention of all the other states and has received total support of the local people. Thus local community participation in the state and their role as watch dogs has been one of the core issues for the success of this campaign. On the other hand the response of the Government and the political will have played a catalytic role in this successful programme.

Why Recycle Waste?

As is well known that there is a regular channel by which recyclable waste materials such as paper, metal, glass and plastics are retrieved from garbage bins and dumpsites by so-called ragpickers, or waste retrievers, who collect the materials, separate them and ensure their passage for recycling. This method of recycling reduces the cost of raw materials while providing employment opportunities to hundreds of people.

In the hilly areas the urban waste disposal option are considerably affected by its specificities like its very fragile ecology and inaccessibility. Landfilling may not be possible due to undulating terrain and paucity of flat spaces. Composting of biodegradable or wet waste may be the predominant choice but with due care to intercept run off from composting areas and its treatment. The recycling of segregated dry waste, therefore, has better feasibility here.

Jan Sewa Ashram: Role of Voluntary Agencies

Jan Sewa Ashram is a leading voluntary organization in the field of solid waste management and recycling through Environmental Education in Himachal Pradesh. It is conducting lectures for Classes VII and VIII in all selected schools about modules for recycling of waste, conservation of natural resources, expected resource crunch and the importance of proper waste disposal. They motivate children of each class to collect segregated waste at home and school. So that it can be transported to the recycling plant. They also established six resource centres/waste recycling centres at the district level for training-cum-demonstration on how waste can be converted into useful and environment-friendly products.

Apart from this programme, JSA and State Council for Science and Technology are running the Secretariat Waste Management scheme in Shimla Secretariat. In this scheme, secretariat waste paper which was earlier burnt is now saved, collected and stored daily. This waste is transferred to a recycling plant after a regular interval where it is recycled into various usable products, which again are supplied to the Secretariat for further use. This scheme is now extended to other 15 government offices. On an average now, in Shimla itself, 4000 kg of waste paper is saved and recycled every month through this scheme.

Under this scheme the shredder company, Methodex Limited has also been involved for shredding the waste paper before recycling. In the second phase of this scheme, waste paper bins make from lantana weed have been installed in individual rooms of this office. The use of lantana bins has been employed as a strategy for the segregation of waste paper as well as benefits to local weaving communities and the control of this weed.

The State Council has organized workshops on the popularization of recycled waste material in all the districts of the State with the assistance of the District Administration to popularize the use of recycled waste products such as paper, plastic bags and agricultural waste materials.

Present Status of JSA in Paper/Plastic Products

JSA through its recycling plant at Jabli, Solan, is preparing all types of paper products of variable qualities. But due to the higher cost of production of handmade recycled paper products against conventional paper mill products, it is reasonably difficult to market such products in the open market without any support. From waste polythene bags, JSA is producing school mats and several other items by weaving strip of polythene bags on the loom. This idea of reusing polybags (against recycling) is the most appropriate and environment-friendly and timely solution of the much publicized problem of waste polythene in hilly areas.

NEED FOR PRICING NATURAL RESOURCES
(NATURAL RESOURCE ACCOUNTING)

This strategy requires two key changes: first, to reform the prices of natural resources and the products made from them (prices that are now fundamentally dishonest because they avoid covering many of the real costs of production) and second, to reform the dominant "world view" that drives modern economics—a view that subordinates nature and the needs of local communities to abstract "national" and "global" economic "growth". The two changes are connected, of course, since the demand for unfettered growth is one of the reasons prices have become so dishonest in the first place, it's easier to show profits when a lot of the true costs are not paid. Prices are blind to most ecological and many social costs. In commodity after commodity critical to the long-term viability of life in the world, market prices are a fraction of true costs. The gasoline burned by the automobile was priced at a lower value as compared to the full cost of that litre of gasoline including such side effects as air pollution damage to human lungs and farm crops (much more than the price). Just about every thing else that uses natural resources intensively is also underpriced.

While laws, regulations and individual efforts to live ethically are crucial in moving toward sustainability, only prices are powerful enough to fundamentally redirect consumption and production patterns. If the prices costs of energy, labor, water, housing, parking spaces and everything else were gradually and predictably aligned, upwards or downwards to match the true costs, and if the money stayed at home, the economy would benefit enormously. Jobs would proliferate even as the environment improved.

Capital refers to physical objects created by people, such as buildings tools and machinery. Land refers somewhat obliquely to all the gifts of nature everything that is not created by people. Land includes not only tracts of earth and natural resource commodities but also basic ecosystem functions, such as the cycles of water, nutrients and energy. These goods are provided by non-human forces, free of charge. The mispricing and consequent misuse of these gifts constitutes much environmental harm.

Why Green Purchasing?

With the emergence of environmental awareness among the consumers and the growing role of green purchasing in the Third World countries, particularly in India, a lot more is needed to be done in changing the way people shop. The issues of environmental benefits of a product and a distinct identity to the environment-friendly products is now urgently required to make the consumers aware.

Moreover, to become self-sustained in future and to continue environment education and waste management, it is necessary to establish a marketing network of recycled products to inculcate the habit of green purchasing "through green shop" of environmentally-friendly products.

The demand for products with recycled content has increased substantially with the rise of government and private procurement programmes which give them preference and experience with recycled-content products. This has also removed consumer apprehension about the suitability in the West. Recycling can be a cheaper disposal method than landfilling or incineration. But a few states are now beginning to make a crucial transition from viewing recycling simply as an environmental measure—a waste disposal strategy—to seeing it simultaneously as an economic development opportunity. Few communities devote the same energy to developing recycling industries that they apply to their collection programmes. The most obvious way to develop markets is to ensure that a guaranteed minimum quantity of goods with recycling potential are purchased. Governments are among the largest buyers of many goods, and among the first prominent market development efforts were state laws requiring or encouraging government procurement of products with recycled content. Twenty-seven states now offer some form of tax incentive for recycling in America. The broad environmental benefits of recycling, especially saving in natural resources and energy will only be realized if manufacturers substitute used materials for a major share of the virgin wood, metals and plastics they now consume.

Compared with producing a ton of paper from virgin wood pulp, the production of one ton of paper from discarded waste paper uses half as much energy and half as much water, it generates 74% less air pollution and 35% less water pollution, saves 17 pulp trees, reduces solid waste going to landfills, and creates five times more jobs. Hopefully, the recycling of paper will soon become mandatory in all countries. A new law in the United States now requires government offices to give preference to suppliers that use recycled fibres in their products. The production of environmentally-friendly paper must take into account the management of forest resources e.g., logging methods and the replanting of trees. The present forestry management practices include large-scale clear-cut logging and aerial herbicide spraying, both of which have a great impact on wildlife and fauna.

What is Required?

Bans and phase-outs are the most effective regulatory tools but require a degree of strong political will and consensus. Economic instruments such as effluent and product charges are easier and less costly to enforce and focus attention on the key elements of production. India is in the process of massive industrial growth, investment in more toxic industries will further impose burdens on India's water, land and air. It is important to ensure that there is a link between liberalization and the environment, invest in a clean production strategy, use a variety of tools including seeking investment in cleaner technologies, utilize economic instruments through command and control legislation.

For example, to drive the expanding commodity polymer industry into manufacturing and marketing less harmful transition materials we suggest that government embraces the principles of Extended Producer Responsibility (EPR). Through mechanisms such as take back legislation, industry will be forced to accept responsibility for its products instead of imposing them upon the community and authorities to regulate. This will lead to manufacturers designing sound ecological principles into products limiting materials and energy use creating true recyclability as well as avoiding the use of toxic substances such as those being generated by chlorine chemistry.

The plastics industry in India is proud of its recycling record. It doesn't want to admit that creating new markets for plastics will lead to waste disposal problems. The principles of recycling waste have always had an instinctive place in Indian society. But it is impossible to apply these to products whose disposal will always create problems. An army of children is ready to pick through municipal or roadside garbage dumps or sit in dark "jhuggi's" sifting through mounds of throwaway consumer plastics, some imported from the north. Out of a population of 940 million, between 44 and 100 million are working children. In badly ventilated workspaces plastics are shredded and compressed into patties and after the addition of dyes extruded and chopped into pellets for making further products. The recycling has two goals, to reduce the amount of new (or virgin) materials and therefore pressure on infinite resources, and to reduce the amount of waste generated by society.

There are several policy tools that the government could use to stem this flow of unnecessary and potentially hazardous products e.g., H.P. Non-biodegradable Garbage Control Act prohibits throwing of plastic waste in public places. The German Ordinance on the Avoidance of Packaging Waste has made manufacturers and distributors responsible for the packaging they create. Its aims were to reduce packaging waste requiring disposal, and to develop sound material use practices. The industry is required to take back materials and either reuse or recycle them, independent of the public waste management system.

The need is to change product material and change consumer behaviour for example Himachal Pradesh Tax on polythene was increased from 8 to 30%. Belgium proposed higher "eco-taxes" on PVC bottles. Even though the law was not adopted, the threat was enough to drive mineral water bottlers into sounder environmental substitutes. The Market for PVC in the agriculture sector in India can be regulated if the Government taxes uses of PVC in irrigation pipes and mulching as this will encourage farmers to use more environmentally sound "transition material" such as Polypropylene (PP) or High Density Polyethylene (HDPE) and also make traditional materials more competitive. The government is not yet facing the plastic waste crisis of the north but wise measures at this point can prevent an unnecessary occurrence of the crisis.

If the external costs of producing PVC were taken into account, then no doubt it would fare badly with traditional materials. This would mean adding the energy costs—pollution cleanup, regulation, and environmental and health damage being done to workers and the community—into the price. Traditional materials, such as stone, wood, bamboo, mud, ceramics, glass are nearly always more suitable and environmentally more sound than plastics. But modern building design often needs light weight materials.

Initiatives Required

It is not a difficult task to change the way people shop, in this media-based and advertised world. From our point of view the following initiatives in this field of Green Purchasing are required:

- Opening of a "green shop" in the main market of the district head-quarters of all six districts with the help of the Municipality, for selling eco-friendly products.
- Exemption of all taxes on the functioning of all such a "green shop".

CONSTRAINTS TO RECYCLING
Economics constraints
Barriers to Buying Recyclables

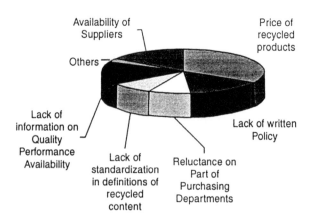

Availability of Suppliers

Price of recycled products

Others

Lack of information on Quality Performance Availability

Lack of standardization in definitions of recycled content

Reluctance on Part of Purchasing Departments

Lack of written Policy

Development of Solid Waste Management Programme in Himachal Pradesh	
Implementing Agency : State Council for Science, Technology and Environment	
ECOLOGY & ENVIRONMENT	NORAD (EPC) PROJECT
1992—Meeting with Kabaries and Rag Pickers	1995-96 and 1997— 12 Waste Recycling Workshops on popularisation of Recycled material held in Districts.
1994—School children involvement through Polythene Cleanliness campaign in School campus.	
	1997—Advertisement of Micro Recycling Units for Paper Recycling, Polythene Reusing & Composting
1995—Polythene Ban proposal sent to MOEF, GOI	
	1997-98—Installation of Micro Recycling Units in districts through:
1995-96—Formulation of H.P. Non-biodegradable Garbage Control Act and Rules.	4—NGOs 8—Schools
Implementation in Shimla Municipal Corporation & Nagar Panchayat, Manali	1996-97-98—Secretariat Waste Paper Management Scheme
1998—Enforcement of Act through :	1997-98—Five Towns Waste Survey & Assessment—Mandi, Solan, Chamba, Nahan and Jawalamukhi.
*State level workshop on Solid Waste Management *Eleven District level Workshops on Polythene Ban	
	1996-97—Financial and Technical assistance in setting up of Bio-conversion Plant in Solan
Polythene Cleanliness drive in Shimla town through ragpickers.	
1999—Strengthening of Act through Prohibition on the use & sale of coloured recycled carry bags	1998-99—State Level Teacher Training Workshops on Waste Recycling
Monitoring of the Act & Implementation of Ban on Polythene bags.	1999—Waste Management Plan for Dharamsala and Manali.

- Promotion of publicity with the help of hoarding and advertisements.
- Exemption of all taxes on propagating the cause of EFP from such a shop.
- A regulating authority at the State and National level for the standardization of quality and market value to EFP (specially recycled products) throughout the country.
- A quality control cell at the state level which can look into the matter of the environmental aspect of a product at its various stages of production and which can classify it on the above basis of whether the product is EFP.

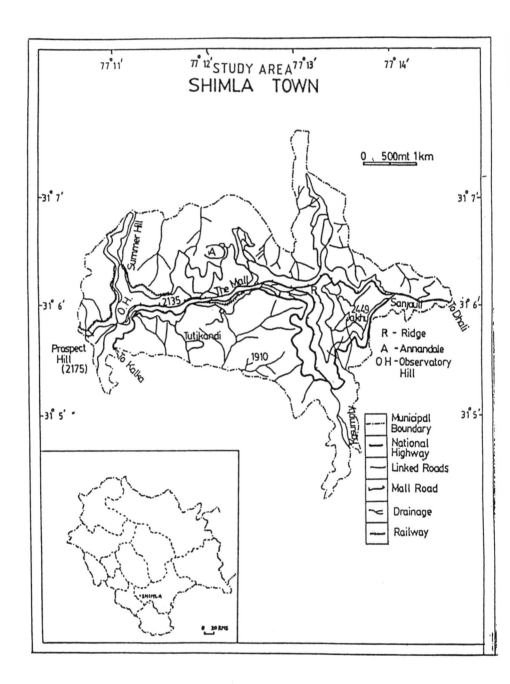

Fig. 1.

GARBAGE GENERATION
(IN GRAMS/PERSON/DAY)

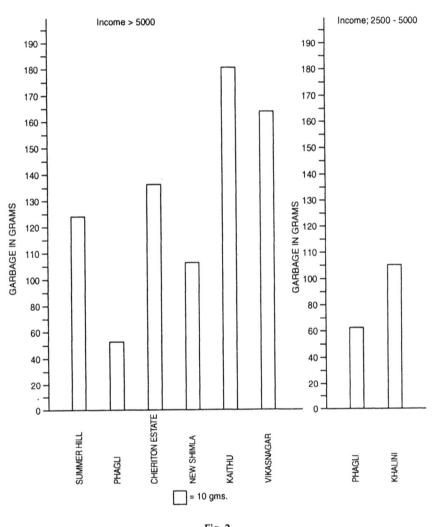

Fig. 2.

- Once a product is classified as EFP there should be no taxes on its marketing, its advertisement and no trade barriers at the national level.
- The establishment of a funding agency which will financially help only those organizations producing EFP for its marketing and trading.

COMPOSITION OF SOLID WASTE
SHIMLA TOWN

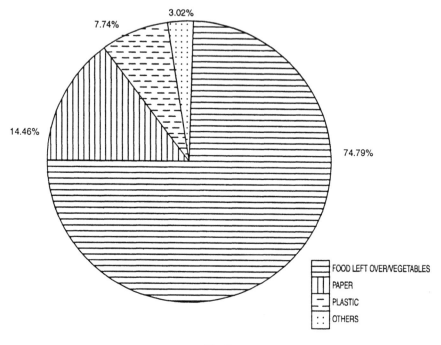

Fig. 3.

CASE STUDY

Solid Waste Management in Shimla Town

The garbage of Shimla town, the capital of the Himachal Pradesh is managed by the Municipal Corporation. Corporation has provided 500 dustbins and forty two dumpers in Shimla town. The main source of garbage generation in Shimla town are house holds, market and offices. The composition of garbage varies from place to place. In market areas, the composition of garbage varies according to the function and services.

Six areas were selected for studying the generation of wastes. A questionnaire was circulated among the residents to know about the generation and disposal of the wastes. The areas which were selected are Summer Hill, Phagli, Cherition Estate, New Shimla, Kaithu and Vikasnagar.

In Shimla town the composition of waste is as follows:

Food & Vegetable lef over   74.79%

Paper	14.46%
Plastic	7.74%
Others	3.03%

The Solid Waste Management team included persons from State Council for Science, Technology & Environment and two NGO's—Vatavaran and Jan Sewa Ashram. Most of the families are well educated and are teachers in the University.

Type	Kg/day
Total waste	30-35
Bio degradable	20-22=(66.6%)
Non-biodegradable	9-12=(30.13%)
Toxic waste	0.5-1=(3%)

Proposed scheme was to segregate waste at source by making households aware of collecting of polythene and papers for recycling to market and also for on site aerobic composting of remaining biodegradable waste.

REFERENCES

Anon. (1993). Report of High Power Committee on Solid Waste Management. Planning Commission, New Delhi.

Beukering, Peiter Van and Sharma, Vinod Kumar. (1996). International Trade & Recycling in Developing Countries : The case of Waste Paper Trade in India. IVM & IGIDR Report, Bombay.

Bijlani, H.U. (1987). Solid Waste Management. Arnold Publishers, Delhi.

Chandel, Jagbir & Kumar, Sanjay (1997). Solid Waste Management : A case study of Shimla town, SCSTE Case Study Project Report, SCSTE Shimla.

Kreith, Frank (1994). Hand Book of Solid Waste Management. McGraw-Hill, USA.

Sharma, V. (1996). Urban Solid Waste Management in Hill Towns. Proceedings of National Congress on Plastic & Environment, FICCI, IPI, Plast India Foundation, Sept. 25-26, 1996, Delhi.

Sharma, V. (1997). Recycling of Waste for Eco-friendly Products: Initiatives and Experiences. Proceeding of National Seminar on Green Purchasing and Eco-labelling, IIMM and MOEF, GOI, New Delhi, 25 April, 1997.

Sharma, V. (1999). Environmental Management Concerns for the city : Some initiatives in Solid Waste Management, Shelter : World Habitat Day Special Issues, HUDCO-HSMI, New Delhi, Vol. 3 & 4, pp. 44-49.

15

The Management of Municipal Solid Waste Using Flexible Systems Approach

Sushil*, R.K. Jalan** and V.K. Srivastava**

*Department of Management Studies,
**Department of Chemical Engineering, Indian Institute of Technology,
Hauz Khas, New Delhi 110 016, India

INTRODUCTION

From time immemorial, human beings have utilized the resources of the earth to support themselves. In early times, the needs were less, resources available in plenty as the population was much less and consequently wastes were not a significant problem. Waste is a result of human activities and has become more prominent during and after the industrial revolution. Today, the accumulation of waste is a consequence of life in an industrialized society (Sushil, 1980, 1984, 1989).

The rapid increase in the density of population in selected areas as a result of urbanization and industrialization is making the collection, treatment and disposal of waste an insurmountable problem, with serious sociological, ecological and economic implications. The limited capability of nature to process the waste on its own poses ecological constraints against which the technological and spatial features of man-made processes in various spheres of activity must be designed and this aspect is gaining importance day-by-day. A sustained effort is, however, needed to restore the socio-ecological balance of nature in order to optimally harness the available resources. It is, therefore, imperative to design waste management systems keeping in view technological, ecological, sociological and economic aspects.

The inefficient and improper methods of waste management, particularly in developing countries like India (Vrat, 1979), are creating pollution problems in the air, water resources and land which are interfering with community life and development. An important feature of waste is its multiplicity characteristic, i.e., uncontrolled or ill-managed waste leads to the creation of more waste, and the failure to salvage and reuse the waste results in avoidable depletion of natural resources.

From the total resources available within a fixed time-frame, a fraction is consumed, a fraction is wasted and the remainder is a reserve. Consequently, the major objective of Waste Management is to minimize the fraction that is wasted.

Naturally, social costs are also incurred for effective waste management and therefore it is essential to carry out a social cost-benefit analysis prior to the implementation of Waste Management Systems. From the above discussion it is clear that waste management is associated with the identification, reduction, storage, collection, transfer and transport, reuse and recycling, and processing and disposal of waste keeping in view health, economics, engineering, conservation, aesthetics and all other environmental conditions involved, in the complete spectrum of the solution to the problem of waste. The present study, however, is restricted to the management of municipal solid waste.

As a consequence of the increasing population of the world, the handling and utilization of municipal solid waste is becoming a problem that is being faced by almost every country. The future scenario looks bleak with more solid waste being generated as the economics of most countries modernize. It is of utmost importance, therefore, to understand and decide as to how the municipal solid waste is to be handled, disposed or utilized economically and with minimum environmental degradation.

SECTORS IN WASTE MANAGEMENT SYSTEMS

In most developed countries, as also in some developing countries, there are basically six sectors to which municipal solid waste can be recycled/ disposed of:

   i)   Biodegradable recycling
   ii) Non-biodegradable recycling
   iii) Composting
   iv) Incineration
   v) Pyrolysis
   vi) Landfills

The sectors of waste management, waste generation and consumption of outputs of waste management are shown in the schematic diagram in Fig. 1.

Recycling of Waste

Recycling of waste involves both biodegradable and non-biodegradable recycling of salvagable waste. Some broad guidelines that could be followed

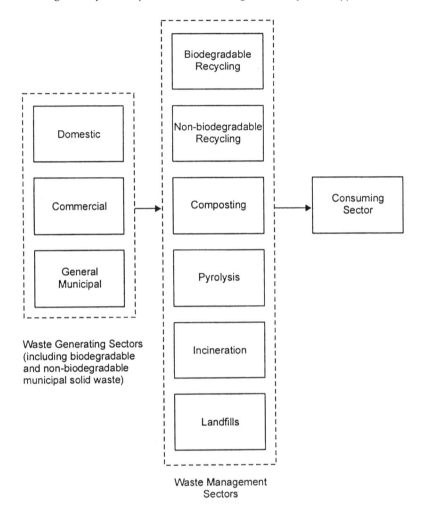

Fig. 1. Schematic Diagram of Overall Waste Management System.

to dispose of salvagable waste include:
i)   Feasibility of recycling to be analysed.
ii) Use waste to produce by-products.
iii) Evaluate feasibility of selling waste/scrap as raw material to other factories/agencies/dealers.
iv) Identify proper processing techniques to generate useful products from the waste.

Recycling is only now becoming important in developed countries. In nature, 99% of everything is recycled (Bellamy, 1993). Nothing goes to waste, and that which goes into long-term storage helps produce the oxygen without which life as we know it would not exist. According to Bellamy, everything must become recyclable and so ultimately the world waste can become extinct. This is not possible until alternative energy source are available, as some recycling procedures are highly energy-intensive.

Plastics recycling has become important as various European countries plan to recycle large quantity of plastics used primarily in packaging. For example:

a) Germany: It is planned to recycle two-thirds of all sales packaging by 1995 comprising primarily of plastics.

b) France: Seventy-five per cent recycling to be achieved by 2003 A.D., including incineration with energy recovery.

c) The European Commission proposed 60% recycling of packaging by the end of this century with 30% minimum for any single material (Snow, 1994).

Waste minimization, however, is more important than recycling for sustainable development (Kersey, 1994). Skinner (1994) feels that recycling or reusing as much waste as possible is important. A major recycling impediment is the question of continued viability and availability of secondary materials markets.

It is important to understand that separation of materials from the solid waste stream by itself does not constitute recycling, which happens only when these materials are incorporated/processed into products that enter the commercial arena (Skinner, 1994).

To analyse the economic feasibility of recycling, one must consider the price received for the recycled material, the solid waste collection and disposal costs avoided and the cost of separating, collecting, and processing the separated materials. While doing the cost comparisons it is essential to consider all environmental costs and benefits. Further, the benefits to future generations in terms of natural resources consumed or landfill space minimized must be considered.

It is essential that waste reduction activities and recycling must be undertaken simultaneously as along with recycling, waste reduction also provides excellent return on investment. Brierley (1993) also indicated that waste minimization at source also leads to payback. Sushil (1990) discussed recycling of waste as an aid to resource conservation.

Based on the above consideration the material-energy flow modules of both biodegradable recycling and non-biodegradable recycling are shown in Figs. 2 and 3 respectively, indicating flow of materials and the waste generation and useful product streams. Recycling has also been analysed in a Systems Dynamic framework (Mashayekhi, 1993).

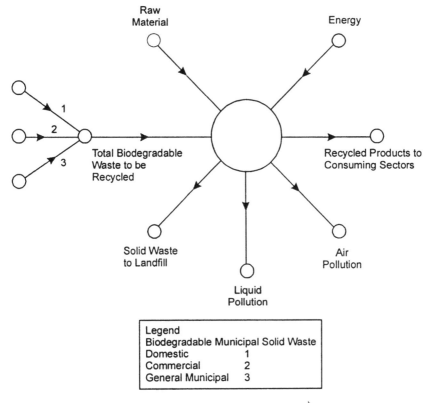

Fig. 2. Material energy flow model—Biodegradable recycling.

The biodegradable recycling sectors have three major inputs of the waste generated: domestic biodegradable recyclable waste, commercial biodegradable waste and general municipal biodegradable recyclable waste. These are mixed and processed using raw materials and energy to produce useful products. For example, while utilizing paper waste, after biodegradable recycling and chemical treatment, high quality paper could be produced after the necessary processing in the paper mill. The effluent generated is the liquid pollutant, the digested sludge is the solid pollutant and the gases generated from the boilers and vapours of the chemicals lead to gaseous pollutants as shown in Fig. 2. Non-biodegradable recycling, however, involves besides other things processing of metallic scrap and includes primarily waste of such types from domestic, commercial and general municipal source. It also includes glass bottles waste, which is processed to produce glass bottles after processing and supply of energy. The effluent and cooling tower blow-downs result in liquid pollution, slag and mud result in solid pollution, gases from the boiler and the furnace,

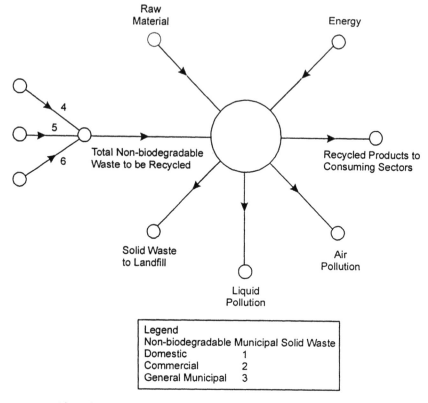

Fig. 3. Material energy flow model—Non-biodegradable recycling.

fumes from the chemicals used in treatment of the scrap lead to air pollution as shown in Fig. 3.

Composting

Composting of solid waste may be defined as the biochemical stabilization of these organic materials to a humus-like substance through scientifically producing and controlling an optimum environment for the process. It is the application of controlled methods that differentiates it from natural rotting, purification or other forms of decomposition. Composting is generally classified on the following basis:

a) Usage of oxygen viz. aerobic (oxygen is used); and anaerobic (oxygen is not used)
b) Temperature: mesophilic (ambient temperature) –20 to 35°C.
c) Technological option
    – Open or windrow—low heap of organic matter
    – Mechanical or enclosed

There are three basic methods of successful composting of botanical and putrescible materials.

1. Open windrowing: This if the first and most common method, because of its basic simplicity and generally low cost.
2. Static aerated pile composting: This second method varies depending on the investment and sophistication of control. This can be either in the open air, enclosed or carried out in specially constructed tunnel chambers.
3. Anaerobic digestion followed by aerobic composting of the residues. This method is now being widely used in the UK.

Finally, the compost is bagged in order to facilitate ease of handling and is essential if retail selling is to be carried out. Bags should be carefully designed as they will have a great impact upon the salvability of the contents. In order to improve the quality of the compost, source separation of municipal solid waste is essential.

Composting in Europe and the USA is continuing to develop successfully and in the UK it is being successfully developed as a method of disposal despite the known low costs associated with conventional landfill. It also has potential in India (Sushil, 1989). It has been used for a long time in rural areas but there is a need to make the people aware of its benefits and a rigorous organized effort is essential. Comprehensive data on the quantity of waste composted in the country is not available.

The flow model of the composting process is given in Fig. 4. Composting includes biodegradable waste from domestic, commercial and general municipal sources which is mixed and processed using bacteria and energy in the presence or absence of air to produce fertilizer/manure used for horticulture/gardening purposes. The percolation of the moisture into the ground leads to liquid pollution; the uncomposted waste leads to solid pollution, the putrefication of waste leads to air pollution and odour formation.

Composting as discussed above could also be considered as a waste disposal recycling technique.

Incineration

Incineration is an important method for waste disposal as it involves thermal processing of waste which results in the reduction in volume and generation of energy and by-products.

Incineration is high temperature oxidation of the organic and toxic compounds present in gaseous wastes is employed to destroy the offensive effluents and odours by converting them into harmless gases, i.e., carbon dioxide and water vapours. Incineration results in 80-90% volume reduction and results in generation of energy.

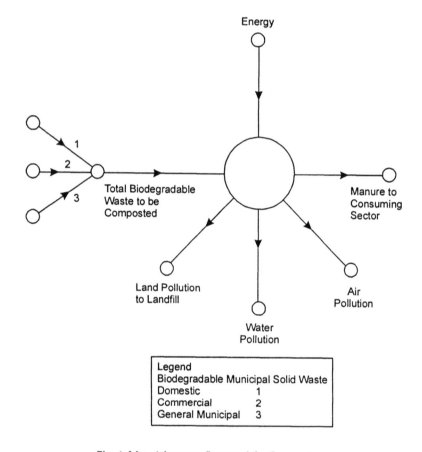

Fig. 4. Material energy flow model—Composting

It is essential to study the following points before designing incineration systems:

- Combustion characteristics of solid waste
- Weight and volume reduction ratio
- Heating/calorific value of solid waste
- Air requirements for combustion
- Trace constituents to control air pollution caused by incineration.

According to Mashayekhi (1993), the incineration sector processes solid waste to produce energy. Incineration produces ash which must be dumped in the landfills. According to Porteous (1993), incineration complements other forms of treatment, rather than being a complete alternative. Incineration is a long-established waste management option and schematic incineration of household waste was pioneered in the UK in the late 19th century Santen (1993). However, it is less cost-effective than recycling and other waste minimization alternatives and it is more expensive than even other waste disposal options, including landfilling. Incineration as a waste management option has been analysed (Bellamy, 1993). Skinner (1994) integrates solid waste management along with incineration.

A material-energy flow model of incineration is shown in Fig. 5 depicting the flow of material wastes involved. In the incineration process all types of waste from domestic, commercial and general municipal sources are mixed of both non-biodegradable and biodegradable type. It is processed utilizing energy leading to the production of more energy depending on the composition of waste, which could be utilized to produce power. The char generated leads to solid pollution which is ultimately disposed of into the landfill. The gases generated during incineration lead to air pollution. The

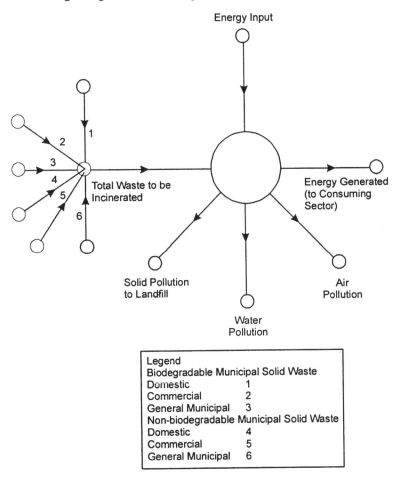

Fig. 5. Material energy flow model—Incineration.

condensed water vapour along with dust and the intermingling of the gases with the water vapour formed during the incineration creates liquid pollution.

Pyrolysis

Pyrolysis is defined as the destructive distillation or the thermal decomposition of wastes in an inert atmosphere producing a mixture of gaseous products, tar, methanol and other organic compounds, water insoluble oils and aqueous solution of acetic acid, and a solid residue, i.e. char is produced.

According to Srivastava and Jalan (1994), the various components produced in pyrolysis are:

Components	% by weight (approx.)
- Gaseous products	25%
- Char	17-25%
- Liquid phase	20-89%

Further, pyrolysis has other advantages which are not present in other waste management sectors such as:

• Production of energy in a convenient usable form.
• Much less air pollution as the process is carried out in an inert atmosphere.
• Recovery of valuable chemicals.
• Reclamation of metallic components.

Pyrolysis is therefore considered to be an industrial process rather than a waste management process. The process of pyrolysis has been studied by Bradbury *et al.* (1979), Brunner and Roberts (1980), Jegers and Klein (1985), Koufpanos *et al.* (1991), Srivastava and Jalan (1994) etc. However, very little work has been done to integrate pyrolysis along with waste management in order to analyse its importance in its entirety.

A material energy flow model of pyrolysis as a waste management sector is given Fig. 6.

The pyrolysis process utilizes the same type of waste as is utilized in the incineration process; however, the processing with supply of energy takes place in the absence of air resulting in formation of solid (charcoal), liquid and gases which could be used as fuel for industrial purposes. The ash generated leads to solid or land pollution. The emissions from the pyrolysis process create air pollution. Liquid pollution is due to the condensation of water vapour which may react with some volatile products, resulting in water pollution.

Landfills

Besides acting as disposal sites for various types of waste, landfills are also

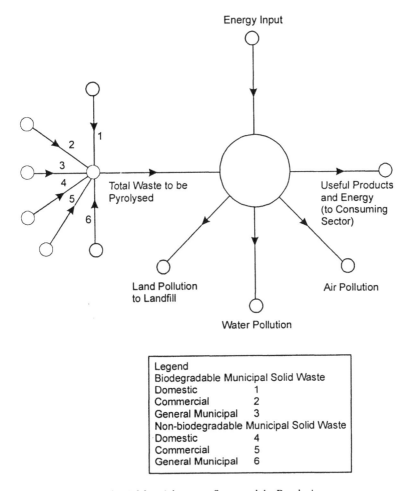

Fig. 6. Material energy flow model—Pyrolysis

considered as an ultimate disposal waste management sector for those solid wastes which are of no use and remain after recycling, processing, etc.

Sanitary landfills have been proclaimed for some time as a low cost, safe way of disposing of municipal solid waste. Wilson (1993) has discussed the development of landfills in an historical perspective in Europe and highlighted the policy and practice with a simple landfill steep ladder. According to Bellamy (1993) landfills are one of the five major methods of waste disposal.

Some of the advantages of the sanitary landfills are:

a)   The initial investment is comparatively low. However, with land prices increasing regularly it is sometimes difficult to acquire adequate land at reasonable prices.

b) A sanitary landfill can receive all types of waste eliminating the necessity of separate collection.

c) It is flexible with reference to quantity and composition of waste being disposed.

d) It helps in the reclamation of low-lying areas.

The above advantages become important provided a proper site is selected for the landfilling operations. The important factors relevant in site selection are:

— Social and political aspects of land use.
— Climate topography and natural features.
— Cost and haulage distance.
— Public acceptance.
— Noise, dust and air pollution.

The common methods of landfilling are: area method, trench method and depression method. Accumulated solid waste in the landfills generate pollution in the environment (Mashayekhi, 1993). However according to Skinner (1994) landfill technology has advanced very rapidly over the past decade. Today's state-of-the-art landfills are equipped with leachate collection systems, liner systems, systems for control of landfill gas, groundwater monitoring, closure and post-closure care, and much more.

The objective is to ensure that landfilling is performed in a manner that greatly reduces the change of environmental degradation, and also that any degradation that occurs is quickly detected and remediated.

However, in the USA the number of landfills continues to decrease. Two main consequences arise:

1) Communities face longer transport distances to deliver their solid waste to disposal sites.

2) Several large facilities, designed to serve a limited number of communities for a given number of years, are seeing their lifespans drastically foreshortened by the influx of waste from outside their service areas.

Crawford and Smith (1985) have dealt in detail with all aspects of landfill technology, and Harris *et al.* (1994) have developed a strategy for the development of sustainable landfill design including leachate treatment by the development of highly efficient bioreactor landfills for mixed wastes, with high leaching rates.

A material energy flow model of a landfill is shown in Fig. 7.

All types of wastes from all sources are mixed and fed to landfills, which are compressed and left below the ground after they are covered to produce methane gas which could be used to produce electrical power or supplied to homes as fuel. In many landfills, a landfill liner of a thermosetting

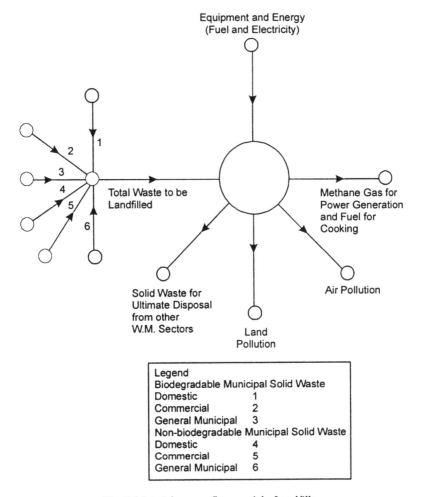

Fig. 7. Material energy flow model—Landfills.

plastic is used in order to reduce the percolation of leachate which is produced in landfills. However, there is leakage of the leachate from the tips of the landfill liner resulting in liquid pollution. The non-converted solid waste, leads to land pollution. The generation of gases other than methane during the process of gas extraction from the landfill which is due to the putrefaction of waste leads to air pollution. Methane gas from landfills is used widely in the UK and other European countries to generate power and also as a fuel for heating of homes and for cooking.

Considering the above, landfilling is perhaps the most important waste disposal method as it also acts as an ultimate disposal to finally process the solid waste.

DEVELOPMENT OF MODELLING METHODOLOGY

Selection of the modelling methodology is based on the conceptualization of the problem being studied and matching the attributes of the problem with the system based techniques. A Physical System Theory (PST) framework integrated with Systems Dynamics (SD) is proposed as an appropriate modelling methodology for the waste management problem for supporting multiple levels of decision making. PST provides a modelling tool for discrete physical flow systems with finite interfaces, which are prevalent in Waste Management. Systems Dynamics, however, provides a framework for studying the system from a dynamic standpoint and with its thrust on the learning paradigm, it would be able to provide more insight into operation of the system. A brief review of Physical Systems Theory (PST) and System Dynamics (SD) is presented and its integration mechanism along with the advantages obtained by the synergistic approach are highlighted.

Flexible Systems Methodology: An Overview

Systems methodologies have been developed as a response to the ever-increasing complexity of systems and the increasing need to adopt system-based methodologies for solving problems with a thrust on implementation in an end-user environment. The task of designing or selecting methodology for a particular problem situation is becoming more difficult with an ever-increasing choice of techniques. The integration of qualitative and quantitative tools is emerging very rapidly to cater to the diverse requirements of management-related problems in general, and Municipal Solid Waste Management, in particular. Keeping the above in view Sushil (1994) suggested a flexible systems methodology, which is built on a spectral, integrative and innovative paradigm. It tries to resolve the end of the continuum paradoxes as it is based on a spectral paradigm, treating all the system-based methodologies on a continuum from hard to soft and also all problem situations from well structured to unstructured.

For example, the Physical System Theory is hard in its design while Delphi, Scenario Building etc., are soft in design. However, system dynamics is somewhere in the middle of the hard and soft continuum. The flexibility in using the systems dynamics methodology has been presented Sushil (1993).

The flexible systems methodology involves the conceptualization of the problem, fuzzy clustering of problem- and system-based techniques, matching of attributes and selection and integration of techniques which is most appropriate for both the problem and solution. This approach has been followed for the selection of the modelling methodology in this present study as problems in real life being considered in this paper are not clustered on the ends of the continuum and lie on the whole continuum and more realistically on the middle part than the ends, having some parts structured and some parts unstructured.

Problem Conceptualization

The problem area has been conceptualized in terms of various attributes and its characteristics as presented in Table 1.

Table 1. Conceptualization of Municipal Solid Waste Management problems.

	Attributes	Characteristics
1.	Nature	Physical flow of waste through various transportation-transformation processes
2.	Structure	Multiple location, flow structure.
3.	Situation-specific characteristics	Mix of stochastic (type of waste) and deterministic processes, reasonable steady state.
4.	Nature of outcome desired	To support decision-making, learning about the waste management systems facilitate policy analysis.
5.	People involved	Not well-versed with quantitatives model good in judgement, wide experience.
6.	Decision environment	Bureaucratic, stress on empirical models, rules of thumb.

From the above conceptualization, it is clear that the Waste Management System has physical flows of items through discrete components with well-defined finite interfaces. It also involves some managerial planning for high level tactical decisions in a multi-location system. In real life the system of waste management is dynamic, interacting with internal as well as external factors.

A fuzzy clustering of the attributes identified above is obtained utilizing the flexible systems approach by considering the various attributes of problem of waste management.

In the system proposed in this study, the problem consists of a mix of operating to strategic problem areas and the decision process involves unitary to pluralistic decision making. Further, the problem situation indicates a mixed character of the operating system and policy decision system indicating that the problem would be somewhere in between the well-structured and unstructured problem, as follows:

Well structured	The waste management problem being analysed	Unstructured

The problem conceptualization indicates that the requirement is for a generalized flow modelling methodology which is capable of covering the entire spectrum of decisions as well as meeting the requirements of multiple levels of decisions. Further, the model must be flexible so that the variable could be incorporated in order to facilitate analysis. In view of the requirements of the problems as mapped on various attributes these two flow modelling methodologies, viz., Physical System Theory and System Dynamics have been identified to have a high possibility of application for analysing waste management systems.

Matching Attributes with System Techniques

In order to match the problem attributes two-system based flow methodologies were selected.

a) Physical System Theory (PST)

b) System Dynamics (SD)

The details of matching of attributes of PST and SD methodologies is shown in Table 2:

Table 2. Matching of problem and techniques attributes.

WM Problem Attributes	PST Attributes	SD Attributes
1. Physical Flow of waste	Physical flow model with material and cost	Flow model of physical flows and feedback of information in levels and storage.
2. Multiple levels involving structured and semi-structured decision making	Structured with focus on deterministic decisions	Unstructured and semi-structured areas with focus on policy analysis.
3. System Dynamic	Analytical solution	Dynamic study through simulation
4. Transparent modelling and graphical representation scheme	Transparent modelling methodology with minimum ambiguity and abstraction	Relatively involved graphical representation.
5. Solution Procedure	Algebraic linear equations can be solved manually and easily	Special Software Package like "DYNAMO" required.

From the above it is clear, that as a flow modelling technique, PST facilitates modelling and analysis of physical flow systems with discrete interfaces having a well-defined structure. PST models have two variables, viz., the physical flow, and the associated cost or energy inputs. The system equations are written as a set of algebraic linear equations and analytical solution is possible. Application of this methodology requires a well structured system and is very well suited for operational decisions.

However, in SD, the flow of information is modelled along with the physical flow, the emphasis being on storages and levels. It is, therefore, used for modelling semi-structured and unstructured problems with focus on policy analysis. SD according to Sushil (1993) is based on the principle that the behaviour of a system is due to its structure and policies. This method permits experimentation with multiple policies and analysis of the effect of these on the system.

Selection and Integration

Matching of the PST and SD attributes with the problem attributes shows that neither of the methodologies by themselves are able to meet the entire set of problem attributes. It is found that PST and SD exhibit complementary attributes in many of the areas and an integration of the PST methodology under a SD framework would meet the problem requirements. Therefore, in order to understand the waste management system, it is essential to integrate the PST and SD methodology and evolve a suitable approach for the WM system. In this system SD is amalgamated with the PST framework as shown in Fig. 8.

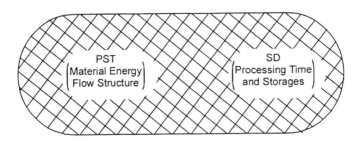

Fig. 8. Amalgamation of Physical System Theory and System Dynamics.

As the waste management problem is analysed using the above methodologies, there is a dynamic shift in the attributes of the problem. The application of the proposed methodology using both PST and SD can be done flexibly depending upon the cognitive burden and expertise available.

The strengths and limitations of the proposed flexible systems methodology for the waste management problem are explained by Sushil (1994) in detail. According to Sushil (1994), the limitations can be overcome by developing a suitable "Expert System" for this purpose.

The flexibility being proposed in this study is systemic flexibility having major attributes of spectral, integrative, interactive, innovative and fuzziness characteristics indicating a trend towards flexibility in waste management.

The continuum approach to Waste Management could also involve treating the various Waste Management sectors ranging from recycling to the ultimate disposal in landfills along a continuum with other sectors falling in the line of the continuum as shown below:

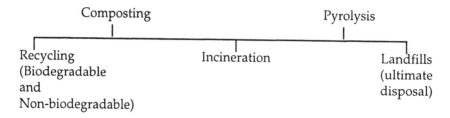

Composting   Pyrolysis

Recycling   Incineration   Landfills
(Biodegradable   (ultimate
and   disposal)
Non-biodegradable)

Here the real situation is considered to be dynamic as the composition of waste changes regularly, is ambiguous and therefore it is proposed to consider a spectrum of it and attempt to integrate, interact, interplay or innovate capturing its fuzziness. Therefore, for managing waste, the flexible systems management is a natural process Sushil (1997).

MODEL DEVELOPMENT USING PHYSICAL SYSTEM THEORY

Physical System Theory (PST) originated in the field of electrical engineering for the analysis of complex electrical networks and circuits. The physical system theory is based on the graph theoretic approach. The graph theoretic approach has flow variables quantity input/output to the system and the costs involved/incurred in the waste management system.

Keeping the above in view, it is possible to utilize the physical system theory approach to waste management. Sushil (1993) has utilized this type of analysis for waste management along with Goal Programming for national planning. Recent work Sushil (1992; 1993) threw some insight into the applicability of PST to various areas. Consequently, this work has been taken to utilize the physical system theory to many critical aspects of waste management and particularly to municipal solid waste management which is a very important problem requiring an immediate solution on a global basis in order to ensure the survival of the planet earth.

As previously discussed, the three major waste-generating sectors as far as municipal solid waste is concerned that are being analysed in the present work are:

—Domestic Waste

—Commercial Waste

—Municipal Waste

Domestic waste is primarily those generated in residential houses. Commercial wastes include those from industrial and office complexes, and municipal waste is that which is generated on the roads, sidewalks, pavements, parks, etc., and is generally collected by the local authority of the area.

Each type of waste discussed above would ultimately by recycled/ disposed of in one or more of the six waste management sectors as mentioned earlier.

Each waste management sector has basically three elements: waste input, output of a useful product/energy and processing of waste leading to further pollution of land, air and water as the case may be, as indicated in Figs. 2-7.

The above concept can be shown schematically in a generalized flow model as given in Fig. 9 for any waste management sector.

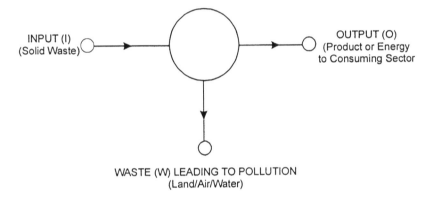

Fig. 9. Generalized flow model of a waste management sector.

In the present study, for all waste management sectors we are considering only one consumption sector.

Based on the overall schematic diagram (Fig. 1) of the waste management systems and the material-energy flow models (Figs. 2 - 7), the cost and flow equations of each type of waste management sector is obtained. The parameters and variables involved are given below:

Parameters

a) Technological coefficients for water pollution, land pollution and air pollution are provided. The technological coefficient is the ratio of pollution generated per unit of waste input to the waste management sector.

b) The technological coefficient for the domestic, commercial and municipal waste is based on the quantity of waste, divided by the total quantity of waste input to the system for both biodegradable and non-biodegradable waste and is computed separately.

Exogenous Variables

a) The cost of processing the waste to produce a product/energy based on one of the six processing systems discussed earlier.
b) The cost of handling biodegradable waste from the source to the composting site for domestic, commercial and municipal wastes.
c) The cost of water, land and air pollution, respectively.

Endogenous Variables

The variables generated while executing the model are:
a) The product/energy output from the waste management sectors model after accounting for waste sectors.
b) The quantity of solid, liquid and air pollution generated.
c) The cost/unit of producing the output/product for the consuming sectors.

The above equations are a set of algebraic equations which are static in nature and can even be solved manually. It is suggested to solve at least one set of equations manually in order to understand them physically before employing a suitable software package to solve the equations on a computer. The Physical System Theory model is integrated subsequently utilizing System Dynamics.

COMPUTER SIMULATION IN SYSTEM DYNAMICS FRAMEWORK

A special package known as 'DYNAMO' has been utilized to solve the above type of equations in a systematic manner so that the flows and cost can be analysed independently. Subsequently System Dynamics had been incorporated by introducing levels and storages through DYNAMO.

Simulation on DYNAMO was done utilizing the cost and flow equations for each waste management sector by characterization of constant and auxiliary equation for each sector. Subsequently SD was introduced by incorporating levels and storages in the model, which bring the processing time and storages in the system. The relationship of flow and state variables describes the dynamic nature of waste management system (Mashayekhi, 1993). Our approach is limited only to study the cost and flow analysis of the physical waste management system in a dynamic mode. This provides the basic model which can then be integrated with other policy issues as discussed (Mashayekhi, 1993).

CONCLUSIONS

The above model could be used for the allocation of waste to different

waste management sectors and could be analysed considering:
—Source segregated waste
—Centrally segregated waste
—Non-segregated waste
The Physical System Theory model integrated with System Dynamics developed for waste management has universal applicability. The input parameters for various conditions depending upon the country, location, and socio-economic characteristics could be altered. The model is simulated on DYNAMO to obtain values for various cost and flow variables and is utilized to ascertain allocation to various types of waste management sectors considering various types of segregation methods by altering the flows to the waste management sectors.

The above model could be utilized for local planning of waste management facilities provided an effort is made to obtain the parameters required in the model, most of which are accessible in developed and industrialized countries. Gradually, such information is also being made available for developing countries. Therefore, it is clear that PST along with SD could be applied to waste management in order to study its causes and effects on the waste management system and understand the dynamics of the entire process involved. PST, integrated with SD, is an excellent tool available to aid planners in the management of waste. Policies of the government, regulatory body, financial considerations, etc., can also be incorporated in the above model in an SD framework. The approach used for planning the waste management system can thus be carried out systematically and flexibly to enhance the effectiveness of these systems. Research work to deal with technical, managerial and economic issues of managing waste, using a flexible systems approach is in an advanced stage to provide a comprehensive framework for developing more effective waste management systems.

REFERENCES

Bellamy, D. 1993. Waste Watching, *Wastes Management Proceedings*, 3-5.
Bradbury, A., Y. Sakai and F. Shafizadeh, 1979. A Kinetic Model for Pyrplysis of Cellulose, *J. Appl. Polym. Sci.*, 23, 3271-3280.
Brierley, M.J. 1993. Environmental objectives: Going for gain not pain, *Wastes Management*, 52-53.
Brunner, P.H. and P.V. Roberts, 1980. The Significance of Heating Rate on Char Yield and Char Properties in the Pyrolysis of Cellulose, *Carbon*, 18, 217-224.
Crawford, J.F. and P.G. Smith, 1985. *Landfill Technology*, Butterworths.
Harris, R.C., K. Knox and N. Walker, 1994. Strategy for the Development of Sustainable Landfill Design, *IWM Proceedings*, 26-29.
Jegers, H.F. and M.T. Klein, 1985. Primary and Secondary Lignin Pyrolysis Reaction Pathways, *Ind. Eng. Chem., Process Des. Dev.*, 24, 173-183.
Kersey, J.D. 1994. Sustainable development and wastes management, *Wastes Management*, 48-52.

Koufopanos C.A., N. Papayannakos, G. Maschio and A. Lucchesi, 1991. Modeling of the Pyrolysis of Biomass Particles: Studies on Kinetics, Thermal and Heat Transfer Effects, *Can. J. Chem. Engg.*, 69 907-915.

Mashayekhi, A.N. 1993. Transition in the New York state solid waste system: A dynamic analysis. *System Dynamics Review*, 9: 1, 23-47.

Porteous, A. 1993. The incineration of waste, *Wastes Management*, **34**.

Santen, A.V. 1993. Incineration: Its role in the UK waste strategy. *Waste Management Proceedings*, 18-23.

Skinner, J. 1994. International progress in solid waste management, *IWM Proceedings*, 3-5.

Snow, B. 1994. Overview of plastics recycling. *Wastes Management*, **39**.

Srivastava, A.K. and R.K. Jalan. 1994. Prediction of concentration in the pyrolysis of biomass materials. Accepted for publication in the *Journal Energy Conversion and Management*.

Srivastava V.K., Sushil and R.K. Jalan, 1994. Development of Mathematical Models for Prediction of Concentration in the Pyrolysis of Biomass Materials, Presented at the *Symposium on Advances in Chemical Engineering* at Indian Institute of Chemical Technology, Hyderabad, India.

Sushil, 1980. *Systems Approach to Waste Management in India*. M.Tech. dissertation, Indian Institute of Technology, Delhi, India.

Sushil, 1984. *Systems Modelling of Waste Management in National Planning*. Ph.D. Thesis, Indian Institute of Technology, Delhi, India.

Sushil, 1989. *Systems Approach in National Planning: A Study in Waste Management*. Anmol Publication, New Delhi, India.

Sushil, 1990. Waste management: A systems perspective, *Special Issue Industrial Management and Date System* 90, 5, 1-67.

Sushil, 1994. Flexible Systems Methodology, *Systems Practice*, 7(6), 633-651.

Sushil, 1993. Application of Physical Systems Theory and Goal Programming to Modeling and Analysis of Waste Management in National Planning, *Int. J. Systems Sci.*, 24, (5), 957-984.

Sushil, 1993. *Systems Dynamics—A Practical Approach for Managerial Problems*, Wiley Eastern Limited, New Delhi.

Sushil, 1997. Flexible System Management: An Evolving Paradigm, Systems Research and Behavioural Science, 14(4), 259-275.

Sushil, 1992. Simplification of Physical System Theory in the Modeling of Manufacturing, Organizational and Other Socio-Economic Systems, *Int. J. Systems Sci.*, 23(4), 531-543.

Vrat, 1989. Waste management in India: The need for organized systems approach, Country Paper, APO Symposium on Waste Management, Japan.

Wilson, D., 1993. The Landfill Stepladder-Landfill Policy and Practice in an Historical Perspective, *Wastes Management*, 24-28.

16

Solid Waste Management in Sri Lanka: A Country Perspective

Ajith de Alwis

Senior Lecturer, Department of Chemical Engineering,
University of Moratuwa, Moratuwa, Sri Lanka

"A Solid waste consultant back from Sri Lanka tells me what he saw when the trucks reach the dumps of Colombo, the capital. The people rush forward first. Then the cows—ahead of the pigs and the goats because they are bigger. Last the crows. A miserable way of life? Undoubtedly ... " (White, 1983).

INTRODUCTION

This quote from 'The National Geographic' perhaps is not the ideal introduction to the topic as it intends to introduce management aspects! Sri Lanka, a tropical island in the Indian Ocean—*The Pearl of the Indian Ocean*—located off the southern coast of India, possesses infinite variety; rich scenic beauty; unusual water, mineral, and biotic resources, and an ancient cultural heritage. These attributes exist today with a population of approximately 18.3 million people in an area of 65,610 sq. km. This rich heritage perhaps today is facing its biggest challenge due to several factors out of which solid waste (mis)management is one.

1) A Brief Country Data Profile:

Area	65,610 sq. km (excluding inland water 64,454 sq. km)
Population	18.3 million
% Urban population	22%
% Rural population	78%
Economic data:	
Per capita GNP	US $ 800
Growth rate of real GDP	5.0%
Administrative structure	

Sri Lanka consists of 25 administrative districts and nine provinces. The hierarchy of regional administrative divisions that supports the central government now consists of provinces, districts, divisions and grama niladhari units, in descending administrative order and area. This is presented in Fig. 1 and indicates the elaborate scheme of government present today. (Fig. 1 also indicates all the cities, towns etc., mentioned in subsequent sections.)

Energy Use

Installed capacity	1526 MWe (Hydro—1135, thermal—391)
Biomass	55% of all energy used
	(others: 27% Petroleum, 18% Hydro)
Hydro	80% of the electrical energy

Environmental Situation

Water quality is the most serious pollution issue in Sri Lanka. The greatest point source is municipal sewage. Only about 17.3% houses have pipe-borne water supplies. It can be stated today, that loss of forest cover, degradation of land, coastal and inland water pollution, vehicular air pollution, and solid waste disposal practices have worsened throughout most of Sri Lanka. Industrial air pollution is still less serious than the vehicular pollution.

Today, Sri Lanka is experiencing the problems due to the mounting 'solid waste'. This situation had been brought about by:

- Better and improving standards of living
- Growth of consumerism (throwaway habits)
- Population growth
- Increasing presence of substances in the municipal waste stream which are difficult to degrade/break down
- Increasing industrial activities
- Poor public participation in finding solutions. People's involvement is normally limited to payment of some indirect taxes at local levels. They, as a group, are more demanding though much less forthcoming with positive contributions.
- Lack of recognition of solid waste as a problem demanding attention and action by the regulatory agencies.

The public is keenly aware of environmental issues and this trend is on the increase. In 1993, Survey Research Lanka, Ltd., conducted an environmental awareness survey. It showed that nation-wide, *environmental problems* ranked fifth (5th) among economic and social problems; However, they ranked *environmental problems*, first (1st) among local community problems for both urban and rural adults (EJF, 1993; NAREPP/IRG, 1992) (Figs. 2a,b,c,d).

Statements such as "the average person has more important problems with which to cope than those affecting his/her environment" and "most

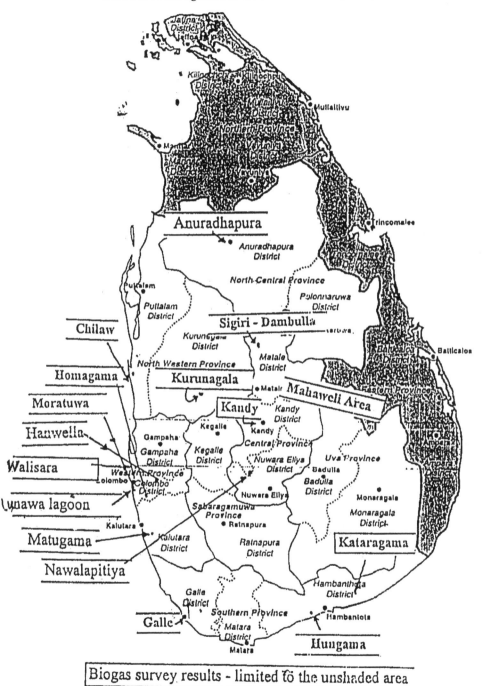

Fig. 1. Sri Lanka.

Fig. 2a. On-going garbage dumping middle of a scenic place, Nuwara Eliya, Sri Lanka.

people do not appreciate the benefits of a clean environment" were greeted affirmatively for 87% and 56% respectively of the population surveyed. Sixty per cent of the sample population is aware of environmental deterioration; more than 80% identified water pollution, *solid waste/garbage,* deforestation, and health-related environmental problems as those currently facing Sri Lanka. It is interesting to note that most times the industry is viewed as a polluter and this is affecting the establishment of new industrial establishments in the country. This situation could be attributed to the prevailing industry practices.

Solid waste from municipal, hospital and industries is a problem that needs serious attention. The issues involved are diverse. For some wastes the non-availability of space for disposal is the main problem. For other wastes, the problem is one of improper disposal. The effects of the latter may range from simple aesthetic consequences to one of safety and health issues involving both human/biological and ecological risks. The magnitude of the problem at present is not huge, though the deteriorating situation is all too common to observe. Thus urgent attention needs to be given to this problem as still, Sri Lanka is operating minus a proper system and standards in place with regard to solid wastes.

Today the solid waste is collected and disposed at a large number of unprotected sites. These sites are not selected based on engineering considerations but more on the basis of convenience or quite simply political

Fig. 2b. Garbage dumping—now abandoned-side of a hill. (Nuwara Eliya)

decisions. However, the problems faced by Sri Lanka are not very much different to those in other developing countries of the region.

Many years ago solid waste was not regarded as a problem in Sri Lanka as urbanisation was not significant and the increasing problems from population densities and crowded cities were not much anticipated. Due to lack of forward anticipation and planning, no action was taken to cater to the future demands that may come with development activities and the rising population along with the resulting urban growth. With increasing population and the proliferation of various industrial activities, the accumulation of solid waste has reached unprecedented proportions. The

Fig. 2c. Indiscriminate dumping of garbage.

problem is most acute in the Colombo Metropolitan Area (CMA) and in other major cities such as Dehiwala-Mt. Lavinia, Moratuwa, Kandy, Galle, etc. Even in remote areas, solid waste dumps have become a common sight. The pressures from urbanization, are compounding the problem. For example Colombo, the capital city, a significant migrant population is added every day, to its population of about 7,00,000. The solid waste load due to this transient population (an additional 5,00,000) is also considerable. This situation is no way in the decline and as such is a problem growing in magnitude with the passing of every single day.

The delay in setting in place a proper solid waste management system, is attributed to the inability of the government to allocate sufficient funds. However, the real issue is not the non-availability of funds, but the inadequate planning and real understanding of the problem and of the possible solutions.

SOLID WASTE GENERATION AND MANAGEMENT

The topic is dealt under three headings considered to be important from a national point of view. Each type has its own characteristics and is important to understand in integrated management planning.
1) Solid waste and sewage from municipal, urban housing and residential schemes

Garbage disposal a 'burning problem': Mayor

Moratuwa group
correspondent

DEHIWALA - Mount
Lavinia Municipal
Council Mayor
Maldeniyage Jayaratne
Peiris said garbage dis-
posal was the "burning
problem" in the area.

"The Municipal
Councils has to transport
about 10 tons of garbage
daily. We have no proper
place to dump this
garbage. But as a tempo-
rary measure we are
dumping the garbage in
a private land in Attidiya.
We have only a few trac-
tors for this purpose. But
we need modern tech-
nology and modern
machinery for the dis-
posal of garbage," the
mayor said.

Call to remove garbage along roads

Panadura group correspondent

The Panadura Urban Council sanitary section says
there is no proper place to dump garbage in the
Panadura Urban Council.

Several years back, the dumping ground at
Malamulla owned by the Panadura Urban Council
was sold.

Urban councillors at several council meetings had
informed the chairman to dispose of the garbage
collected in every road in the Panadura Urban
Council area.

Dirty stories of Colombo city

Rs. 500m worth garbage clearing equipment for Colombo arrives

18/03/97 (By Wyman Hettiarachchi)

A NEW fleet of 49 garbage collecting trucks and
accessories valued at Rs. 500 million ordered by the
Colombo Municipal Council from Japan during Mr. K.
Ganeshalingam's tenure as Mayor has arrived in
Colombo.

Mr. Ganeshalingam who found that garbage collection in the Colombo city was severely hampered due to a
shortage of clearing trucks contracted to import an
additional fleet under Japanese aid and scheme. Chandrika
Bandaranaike Kumaratunga and 'got her
waived the import duty on these vehicles.

A municipal spokesman said the council currently
had a fleet of only 55 vehicles of which 12 were imported
after Mr Ganeshalingam became Mayor.

This fleet was grossly inadequate to clear nearly 700 to
750 tons of garbage heaping up in the city almost daily.
The result was that nearly 100 to 150 tons of garbage
remained uncleared in the city daily.

The spokesman said the new vehicles would enable
the council to clear the city of all garbage daily.

A workforce of 2200 persons will be deployed in the
city to collect garbage at the clearing points.

A 40,000 sq. ft. shed has also been erected at Deans
Road. Maradana to serve as a transit point for the
garbage so collected.

Dirty deal on dumping grounds?

By M. Ismeth

(Moratuwa dumping ground adjoining
complains that the stench emanating from
here is a dumping ground ...)

Colombo city in the 200 acre
sit at Welisara. But officials ...
tipped officials were now being
on the project site at ...
project ...

The official referring the
large cumbersome loads of ...
Kelaniya proposed bidders for this tender
... the only bidder for this ...
the Rs. Treasury studies of a ...
two years ago said ...
documents and modern technology ...

An official of a Welisara disposal filling World Bank multi-
lion truck project that ...
land ...

Garbage dump at Lotus Road

Lotus Road in the heart of Colombo Fort was once
a gateway to the city when the passenger jetty was
located opposite the P & O hotel. It was then a
clean and a well maintained road. That was the time
Colombo was considered a clean city.

But all that old fame is now vanished. We cannot
now boast of a clean city.

It is now a common sight in the city to see heaps of
garbage accumulated everywhere and not removed
for days. Lotus Road is no exception. The garbage
dump at the Main Street end of this road opposite the
YMBA building, where a cafeteria is also being
run.

Garbage dumped here is scattered and not removed
for days. This is an environmental and a health
hazard and a paradise for stray dogs and crows.
The waste water that runs down this stretch of road
are parked close to the garbage dump. Buses on Route No 170
commuters are compelled to wade through the filthy
water to get into a bus.

The CTB time keepers old hut which is hardly
used (there are no CTB buses here now) is another
eye-sore, which should be removed.

The Colombo municipal authorities should take
immediate action and ensure a clean Lotus Road by
removing the garbage collected here at least once a
day without allowing it to rot.

Fig. 2d.

ST 97/05/30

A dumping ground

Lewis Place and Porutota Road are the main areas where you find most of the tourist hotels, restaurants, gem and jewellery shops and other establishments that cater to the needs of the tourists who visit Negombo.

It is regretted to find that this whole area - Lewis Place and Porutota Road - is like a dumping ground for garbage, dirt and other waste matter. You just can't walk on the pavements (in fact there are no pavements at all!). Garbage is dumped on both sides of the road. Besides roads are dug up by the Telecom, Water & Drainage B... Electricity Co. all the year round and their work never seems to be complete. The diggings are just covered up without levelling or tarring. There are plenty of pot holes everywhere like moon craters and when it rains the water has no way of running out. One can hardly walk on the roads for they are full of potholes and stones left for good. Motorists find it very difficult to drive owing to the bad and grave conditions of th...

Stray cattle ... other ... spared. You will find it smelling of human excreta and heaps of cow-dung along the beach.

The street lights opposite the Browns Beach Hotel are not lit up or are very dim and the tourists fear to go for a walk or for shopping for fear of petty thieves, pimps ... other unwanted char... drug addicts ... touri...

June 5 was World Environment Day

People doing their own thing in paradise

Legal action against Kalutara UC for polluting area

Wadduwa special correspondent

THE Kalutara Pradeshiya Sabha Chairman Gamini S. Kaviratne, said he will file a case against the Kalutara Urban Council for polluting the environment by dumping garbage at Pohorawatta, Nagoda, a land belonging to the sabha.

He said so following a complaint made by Mr. M. G. Perera (SLFP) that the area was being polluted due to the council dumping the garbage at Pohorawatta.

"The waste matter is washed into the nearly paddy fields creating a major problem to the farmers and the tax-payers. It has become an ideal ground for breeding mosquitoes", he said. At the sabha's monthly meeting held at Kalutara recently.

DN / Friday / Jan. 10, 97

Pollution of the Environment

I am not a frequent visitor to Nuwara Eliya. But on the last two occasions I have taken the route via Avissawella and was very impressed with the ... until I reached the ... Eliya just ... not appear to even know the basics about hygiene to dump garbage on a slope of a stream and allow it to fall into the stream. The stream water would carry the bacteria right through the country. Surely, can they not find a better method of disposing of rubbish?

Zarook Marikkar

Mr. Aboosaly of the Hatton-Dickoya Urban Council says that this site has been used for dumping garbage since the inception of the Urban council in 1931. During the last two years the U.C. has taken steps to obtain an alternate site for dumping garbage. To obtain a suitable site for dumping ... on the April ... Aboosally says the Council will work out ... very soon this site will be cleared and the garbage dumped in another ...

A teacher - St.Joseph College - Bandarawela - 28.

Environment is a very wide term. We say that the success of a ... dependent on a good social environment. I gear my children to a ... there is a good social environment. Then we say a religion e... necessary. We should be able to create a good and healthy c... people. Let us forget the social environment and the religious envir... focus our attention on natural environment conversation with a Buddhist priest at Matale, Kaludewela. Still the Town Council dumps the waste near the river. I have fought hard against this. Even the waste of the slaughtered cattle is dumped into the river. I have fought to stop this pollution of the river. I did a hunger strike for 8 days. Having fasted for 8 days, I thought if self-immolation on the last day and was ready with petroleum. The only weapons I had was pen and paper. What I wrote was published in the papers both in Sinhala and English papers; Divayina, Dawasa, Daily News, Sun - People bathe in this river. The past we used this water for drinking. The night soil collected by the Town Council was earlier buried in a pit. Later they started dumping that too into the river. Even the dead cattle are thrown into the river.

... medical clinic than a dirty unswept ... webbed ... with clogged drains ... high heaven. Every resident knows that ... in spite of rates and taxes ... and buildings being ... sucked up.

Of course they can trot out a multitude of reasons for this neglect - that is a national pastime - but the people are interested in jobs that are done ... done if there is a will.

It is time that the central government sat up and took note of this pathetic state of affairs. There are ample grounds for dissolving ... of the councils mentioned at the beginning of this article.

Special Commissioners are better suited for the administration of these areas under the present circumstances where people's confidence is eroded and hostility is mounting. They should be men ... of integrity, ability, young ... energetic, disciplined and with ... party affiliation. Not the desk bound type ... super annuated administrators still struggling to survive fawning on anybody and everybody in power, knowing the way from the garage to the air-conditioned room, but those who will frequently go to the field to learn at first hand what is happening or not happening.

The ball is in the court of the Central Government. Will it act or merely turn a blind eye?

Fig. 2d. Contd.

Letters to the Editor

Industrial pollution in the suburbs

Industrial pollution and protection of the environment are the major topics that are discussed by high officials and five star hotels by high official in protecting the envi...

Either ...

Moratuwa: Black smoke causes residents to flee

Council had a few months ago begun dumping garbage in the water logged bags, in the land saying it wanted to fill up the land and sell it. For the past few months, residents said, they could not bear the stench. Three days ago the garbage caught fire. Fire Brigade came to put the fire off, Colombo but the fire could but left saying they and chest pains.

Several families in the residential area, off the Moratuwa-Egoda road with their friends have abandoned their no longer homes as a result palatial black two acre of a thick smoke, a two acre smoke from their garbage dumping ground.

Residents of the area Morauuwa Urban occasions when children refuse to said the Moratuwa to the bad smell. Recently due to some chemicals dumped in this yard, the whole yard

fire. It was covered with a thick dark smoke. People rushed their children to safer places and fighting the fire with water. With all these people around this area are fighting for which I believe no one knows.

polluted area one should notice is located College Avenue, Mt. Lavinia. Every day people, locals and tourists come here for will find garbage dumped on both way line. Also toilet waste and toilet ...ls around the area are drained to the

not douse the fire as there was inflammable material, and polythene, nearby like from garbage chemicals in the factories in the heap.

Upto yesterday the fire was still smouldering and emitting smoke which caused black breathlessness and severe coughing ...and old ...orities must consider it their duty to prevent environ-

...ple are unaware of this and they How many may have got virus diseases only god knows. Garbage that is ...ly kitchen waste from hotels and ...smell ... ...

...c environment for us and for our ...ure generations is our responsibility and the auth-

...mental pollution.

RANJITH BANDARA
Kohuwala.

MUNICIPAL WASTE COLLECTION POINT

Fig. 2d. Contd.

Crab and Shrimp Shells for Cash

In Sri Lanka, food users of shrimp and crab are having second thoughts whether to discard the shells or retain them. This came about following reports that shells from these seafoods rich in a natural polymer called chitin and can be recycled for cash. Chitin has many industrial applications such as purifying water, increasing crop yield and healing burns and wounds. Food processing plants use chitin for treating polluted water.

Usually, 8 tons dried shell give 1 ton of pure chitin, which sells at US $8 per pound.

2)   Industrial waste including hazardous waste
3)   Hospital (infectious [healthcare]) waste

1) Solid Waste and Sewage from Municipal and Urban Housing and Residential Schemes

The most significant contribution to solid waste arises from this source. The quantities of solid waste generated in the island are not well quantified. The best data are available for the Colombo City and the metropolitan area (CMA). They have partly risen from numerous studies done to initiate a solid waste management system to the city of Colombo. Colombo municipal area is producing about 750 metric tons daily and the figure for the whole metropolitan area is about 1,000-1,100 MT/day. The per capita solid waste generation figure is 0.98 kg/day. The per capita generation of solid wastes has increased from an average of 0.75 kg per person per day in 1986 to about 1.0 kg per person per day. The expected value in the year 2010 is 1.25 kg/person/day. Ninety per cent of the solid waste generated within the CMA is collected for disposal while smaller local authorities collect as little as 5%.

Sri Lankan environmental regulations (The National Environmental Act) require that only if more than 100 MT/day waste is to be dumped at a particular site, does the site then needs to undergo an environmental impact assessment (EIA). This means, that as outside the CMA the quantities of solid wastes generated will never exceed this value, the municipalities and local authorities need not subject their action to an environmental assessment process and can escape from a detailed environmental scrutiny.

The waste composition has been studied by various study groups (JCI, 1981; ERM, 1994a). The results can be broadly summarized as follows (Tables 1, 2). The composition indicates that about 85% of the waste is organic and has a moisture content of about 60-75%. The calorific value of this waste is considered to be about 600-1000 cal/g. These data have been largely determined for the waste arising in the Colombo area. The distribution can, however, be taken to be comparable with other regions where the change perhaps will be to have a higher organic fraction.

This shows the predominant component in solid waste to be the organic fraction. As such discarding this component by considering it purely as unwanted waste means wasting away a valuable resource.

In Sri Lanka, only Colombo boasts of a sewer system of any significance. The Colombo system serves only about 482,000 people, whereas other residents rely on on-site disposal (septic tanks). It is estimated that about 15% of the total population of the Colombo urban area have no sanitation facilities at all. The country is yet to have a municipal wastewater treatment facility. Colombo sewerage system, which is quite old (i.e. 1902), basically

Table 1. Waste composition data from published studies

Study Year	Study 1 1980		Study 2 1981			Study 3 1993			
Material category	All resid-ential	Com-mercial	Low-income	Middle and high-income	Com-mercial	All resid-ential	Low-income	Middle and high-income	Com-mercial
Paper	8.00	28.00	7.00	11.8	0.4	7.02	5.63	7.48	6.73
Plastics	1.00	1.00	1.1	2.8	0.0	5.62	5.37	5.7	5.2
Metals	1.0	1.0	2.0	5.0	0.0	2.03	0.64	2.5	0.63
Glass	6.0	8.0	1.9	2.8	0.0	0.66	0.0	0.88	0.0
Organics	82.0	61.0	79.1	64.7	92.8	83.58	85.3	83.0	86.4
Others	2.0	1.0	8.9	12.9	6.8	1.09	3.02	0.44	1.0
Total	100	100	100	100	100	100	100	100	100

Table 2. Variance in refuse composition by source of generation, Colombo (in percentage by weight).

Type of material	Residential	Market	Commercial
Paper	8	8	28
Glass, ceramics	6	<1	8
Metals	1	< 1	1
Plastics	1	< 1	1
Leather, rubber	-	-	-
Textiles	1	1	1
Wood, bones, straw	1	0	2
Non-food Total	18	10	41
Vegetative, putrescible	80	88	58
Miscellaneous inerts	1	2	1
Compostable total	81	90	59
TOTAL*	99	100	100

* Total = (Non-food Total + compostable total)

consists of a reticulation system leading to pumping stations, which are linked up together with a set of force mains and a few gravity trunk mains. The system is divided into a northern and southern system the collections of which are discharged to the sea through two sea out-falls (about 2 km into the sea).

There is scarcity of land in and around Colombo. Low lying lands are a frequent choice as disposal sites which in addition to being unsuitable due to pollution effects, tend to reduce flood retention areas thereby aggravating an already bad flood situation. The Beira lake, Lunawa lagoon, and the Kelani river are the main water bodies being polluted through the uncontrolled and unmonitored discharge of domestic, commercial and industrial sewage and waste.

Local authorities have been resorting to the practice of dumping garbage at certain selected locations without any consideration to the adverse effects of such dumping. The sites are solicited by placing local advertisement or informally through contacts. Most people rent out land to get them filled so that at the end, the filled-up land can be covered with soil and sold at a higher price. These activities put neighbours into lot of inconveniences due to bad odour, dust, animals and pests, vehicular noise and associated health hazards. In these situations when the dumping is going on, the garbage is not even covered with a layer of earth. In the hill country, the practice has been to dispose the garbage at a hillside away from city centres. The leachate from these sites often ends up in water bodies polluting them significantly, as the leachate is very high in BOD and COD.

The effects of improper disposal practices are affecting Sri Lankan scenic sites as well (Tantrigama, 1995). This is also a serious issue as tourism is an important foreign exchange earner. There are instances where visitors, especially those from overseas complain about the deterioration that has taken place over time. Garbage disposal facilities are inadequate at tourist sites, which is one of the main reasons that garbage is thrown everywhere by visitors. In festive seasons the problem becomes worse. The responsibility lies mainly with the relevant local authorities for initiating schemes and establishing maintenance mechanisms. The objective should be to maximize reuse and minimize disposal needs. Public participation again is vital, but with proper facilities should not be difficult to obtain, as in these places visitors always understand the intrinsic value of the place.

The two main options available in utilizing the organic fraction without involving much capital or expertise, i.e. composting and anaerobic digestion are not practiced in Sri Lanka to the extent desired. The awareness of these techniques is quite limited. There is only some minimal activity in this area.

a) Anaerobic Digestion (Biogas generation)
Biogasification (Anaerobic digestion) was introduced to Sri Lanka in 1969. The energy crisis of 1973 prompted more activities in the area though there had not been activities to the same magnitude as in neighbouring India, Nepal or as in China. As an agricultural-based country, significant quantities of wastes are generated of plant and animal origin. Animal husbandry is

practised widely in the country as well. The biogas systems used are the conventional low-rate digesters.

De Alwis (1996) revealed the following distribution of biogas system use in Sri Lanka after an island-wide survey. The survey was not carried out in the districts of North and East due to the prevailing political situation. Table 3 gives the summary of this survey on biogas systems. These systems predominantly use animal wastes (from cattle, pigs, etc.) followed by straw and human waste. Straw is used in a system developed by the National Engineering Research and Design Centre (NERD) and is called a Drybatch (Fig. 3a,b,c,d) system (Anon, 1991). This system was awarded a silver medal in an exposition held in Geneva on environmentally-friendly technologies in 1996. Drybatch system, as its name implies, is a batch system and does not require water in any significant way compared to systems such as Chinese units where usually a ratio of 1:1 (vol:vol) :: Solid waste:water is followed.

Table 3. Biogas Systems in Sri Lanka: National Survey Summary, 1996.

District Summary				
District	Total Surveyed	Working Systems	Under Construction	%
Colombo	19	11	02	64.7
Gampaha	44	19	00	43.2
Kalutara	03	02	00	66.7
Anuradhapura	02	02	00	100
Kandy	70	26	02	33.8
Matale	13	02	00	15.4
Nuwara-Eliya	07	03	00	42.9
Galle	05	02	00	40
Matara	07	01	00	14.3
Hambantota	58	01	00	1.7
Puttalam	16	07	00	43.8
Kurunagala	102	20	00	19.6
Badulla	05	01	00	20
Ratnapura	07	02	00	28.6
Ampara	06	04	00	66.7
Moneragala	01	00	00	00
Kegalle	04	04	00	100
Total	369	104	04	701.4

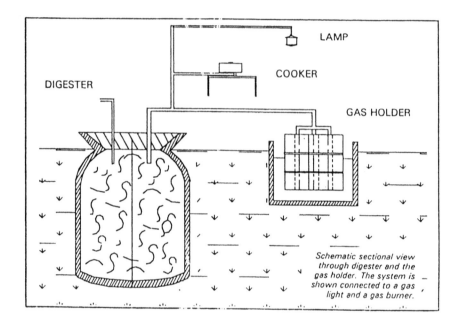

DRYBATCH digester system at NERD—using 700 tons of garbage (Market waste)

Fig. 3a. Drybatch biogas digester system.

Cumulative Summary

Total units surveyed	369
Under construction (observed)	04
Systems surveyed (complete)	365
Functioning units	104
Functioning rate	**28.5%**
No. of units abandoned after successful use	16
Success rate	**32.9%**

Type of digester	% of Total		Functional units	Efficiency
No. of Chinese units	313	85.8%	95	30.4%
No. of Drybatch units	23	6.3%	06	26.1%
No. of Indian units	27	2.7%	00	0%
No. of other units (Septic)	02	0.55%	02	100%
Digester setting	**% of Total**		**Functional systems**	**Efficiency**
No. of Home systems	278	76.2%	81	29.1%
No. of Farm systems	22	6.03%	7	31.8%
No. of Institutional systems	65	24.5%	16	24.6%

Fig. 3b. Drybatch small unit at the NERD Centre digester in front barrel collections behind.

The cumulative figure indicates that only 28.5% of the systems surveyed were functioning *(Functioning rate)*. An earlier 1986 study indicated that 61% of the units were functioning. However, if one to consider that in the Hambantota district in the Hungama village the units had served their useful intended purpose (16 units) and that the number now not functioning to have been successful the *Success rate* can be said to be 32.9%. Not all units from Hungama had been considered in this as some units had shown technical faults from day 1 itself! Similar considerations in other districts may slightly improve the success rate.

Fig. 3c. 600 ton digester pilot plant NERD Centre, Colombo for municipal market garbage (note lighting panel using the gas in front).

These summarized results indicate that the *success rate* is low (32.9%). From the technologies used, Chinese systems running with animal waste (85.8%) appear to be predominant with the majority of the systems based in homes (76.2%). The drybatch system introduced by the NERD centre is not known in areas where it can be most effective (i.e., Mahaweli area and in the dry zone where paddy is cultivated and straw is generated as waste). It also appears that the use of biogas in farms where it can play a significant role is not taking place to the desired extent in Sri Lanka.

Nawalapitiya urban council has implemented two batch type biogas digesters (60 and 30 ton units) to digest market waste. The biogas generated however, is not utilized to the full extent possible, although at the onset a nearby bakery had been earmarked to use this gas. This has happened mainly due to the lack of interest and support being given to this pilot project. Being a batch unit, a number of units are necessary to manage the waste generated daily. Recently the Colombo Municipal Council has also initiated a pilot project on biogas generation using market waste. The technology used is that developed by the NERD centre (Drybatch method). Prior to this project NERD demonstrated the feasibility via the setting up of a 700-ton digester unit.

Some developments are presently underway for Colombo Municipal Council, Maharagama Pradeshiya Sabha and Anuradhapura Municipal Council areas in trying to utilise the garbage in these Drybatch systems.

Fig. 3d. Kirulapare market garbage digestion unit for Colombo market waste (Developed by the NERD Centre baud as their drybatch concept).

For Colombo, the project is being done at Kirulapone with construction of 05 digesters of 40 MT capacity. These are to handle the Colombo Manning market garbage which is about 7-8 tons per date. To solve the problem of market garbage at Maharagama it has been calculated that 16 digesters, each of which has sufficient capacity to accommodate one week's garbage in the area (i.e. 10 MT capacity), are necessary. The expected cost in this instance is about Rs 0.21 million (1 SLR = 65 US $) per digester. This means that the area required and cost component is significant. This is an indication of the shortcoming of adopting low-rate systems to handle large-scale problems.

The Kirulapone 'Market Garbage Anaerobic Digester' scheme (4 units at each with 40 MT capacity) became operational with a ceremonial handing over of the facility by NERD centre to the Colombo Municipal Council on 18th Nov. 1998. The gas collected is used for running a diesel generator of 5 kW and for boiling water during day time. As these units are run in batch mode the market waste generated for three weeks will occupy these units for approximately 4 months. Labor is involved in waste removal and also in digester handling and this is inconvenient as the work involved is not pleasant and easy. The main emphasis at this stage with these units is from the sale of soil-conditioner, which is expected to generate much revenue.

It is still questionable however, without resorting to some high-rate anaerobic digester systems to economically utilise the organic substrate fraction of the garbage as the kinetics of these drybatch systems are low and thus require large number of units to effectively deal with the quantities produced.

b) Composting

The option of composting perhaps is the most appropriate in the present climate as the method needs less attention and the systems can be expected to perform more by default. The compost generated can serve a useful function in providing a useful soil conditioner. There is general understanding about the usefulness of this method. Even vermi- composting has been promoted at times. The Ministry of Agriculture tried to promote small-scale composting of municipal wastes in low income urban areas in 1981. Circulation of information leaflets to encourage small-scale composting of municipal wastes in low in-come urban areas took place during this time (Sathianaman, 1981).

CEIP (Colombo Environmental Improvement Project) is keen on promoting composting with private sector involvement. A market survey conducted in 1993 to determine the potential market for compost, indicated that the plantation sector represents a potential demand of approximately 180,000 MT/year. At present there is one private sector company involved in composting. The production facility is at Anuradhapura at the heart of the agricultural area of the country (i.e., in paddy). The only local authority to have used composting is the Matugama Pradeshiya Sabha, which has now abandoned the project.

Presently work is being done at University of Peradeniya (Fernando and Basnayake, 1998) on developing a process of composting for the Kandy Municipal Council. The pilot scale work had been successfully completed and the project is at the stage of being promoted at local council level within the Kandy district. In addition the Urban Development Sector project funded by the Asian Development Bank and managed by the UDA, also has initiated as a sub-project the use of composting to handle urban waste (Figure 4).

c) Incineration

The incineration option has been the most suggested option in the management of solid waste. The suggestion has not been made with any real understanding as the composition and the calorific value suggests the unsuitability of incineration as the management option. There is not a single proper incinerator system working in the country. Incineration perhaps looks attractive as there is not much to do except initiating a fire and those who suggest this option anticipate the solid waste bulk disappearing away into thin air. No local authority will be able to support incinerator systems with additional fuel demands that will be made to keep this type of low calorific value, high moisture content waste being properly burnt.

Ministry of Health is at present has identified incineration as the option in the management of healthcare waste and is in the process of acquiring an incinerator initially to deal with the waste generated from the major hospitals in and around Colombo.

2) Industrial Waste

Sri Lanka does not have a significant industrial waste as our industrial base is not strong. However, in the limited number of industrial activities there is gross mismanagement of solid waste generated. Some examples are:

Moratuwa city which is about 17 km from Colombo is famous as the furniture capital of Sri Lanka. There are over 1100 saw mills and furniture manufacturers in operation in the Moratuwa area. The majority of these concerns do not have an environmentally-sensitive method for disposing their saw dust. Although this waste is a valuable energy resource, it is not utilized as such. The majority of this saw dust is being dumped into the regions lagoons, lakes and rivers. The saw dust is being deposited at such a rate that it has no time to degrade, and is clogging waterways, causing untold damage to the environment. This has directly contributed to the flood problem in the area. Although now an illegal activity, and subjected to a fine, it is still widely practised as there is no other economic alternative available.

Moratuwa-Ratmalana area is also an industrialized zone. The area has about 300 factories of various types i.e., paints and coatings, food processing, asbestos, metal finishing and fabricating, textile processing, glass manufacture, machinery maintenance, battery manufacture, etc. These industries have not developed in a systematic manner and there is considerable pressure between the local residents and the industrialists. Most of these industries do not have any wastewater treatment units and a common method of solid waste disposal are either washing the waste

Fig. 4. An on-going construction of an open composting facility for Nawara Eliya city.

with lots of water, burning or illegal dumping. For instance waste from the asbestos facility ends up being dumped on a low-lying land (Fig. 2b).

Industry is heavily concentrated in Colombo and Gampaha districts, which includes more than 40% of the country's private sector and more than 60% of public sector establishments. Industries, such as tanneries, which are scattered around the capital city have acquired a bad reputation due to the current disposal practices of solid and liquid waste. The government has now approved removing all tanneries from their present locations and placing them in an area with a centralized treatment system away from Colombo.

Most of the industrial waste generated ends up in the normal solid waste dumps used by the communities. In some situations the industries sell containers, barrels etc. filled with quantities of waste. The buyer, being only interested in the container disposes the waste contained in the most convenient manner available, i.e., dumping. The laws to prevent such practices are being tightened with amendments being incorporated to the National Environmental Act. Few industries have incinerators to burn solid waste though they too fall below the normal criteria of acceptance. The country is yet to gazette air source emission standards and as such the industries cannot be made to improve on their performance.

Away from Colombo, in the districts of Anuradhapura and Pollonnaruwa, the paddy mills discharge their solid waste 'paddy husk' to waterways causing sedimentation and water pollution. Similarly, in the coconut-processing belt Chilaw, Kurunagala areas, coir dust from coconut processing, is a problem.

a) Hazardous Waste
The Government of Sri Lanka in 1996 carried out a pre-feasibility study on hazardous waste management and disposal for Sri Lanka (ERM, 1996). This study involved an estimation of the quantum of hazardous waste generated in Sri Lanka as this would indicate the nature of the existing problem. The wastes considered were mainly solid/semi-solid/liquid hazardous waste-streams from industries. The survey estimated that at least 40,000 tons per annum of hazardous wastes are generated in the Colombo metropolitan region (Colombo, Gampaha and Kalutara districts). This comprises at least 10,000 tons of inorganic wastes, 14,500 tons organic wastes and 14,000 tons of oil waste from motor vehicles. Except that of 14,000 oil waste of which a significant percentage will end up as a secondary fuel, the rest is destined for dumping or ad hoc disposal. As most industrial activities liable to generate hazardous waste are concentrated in the districts of Colombo and Gampaha, the value will not exceed significantly if an island wide survey had been carried out. It is common to see factories disposing waste into open dumps along with other waste and as such hazardous waste also ends up in common dumps.

3) Hospital Waste

The waste generated from hospitals does not yet receive special attention although their hazardous and infectious nature is understood. Some hospital waste is flushed into the sewerage system and, therefore does not enter the solid waste stream. Hospital waste generated, are estimated currently to be in the region of 5 tons per day, of which less than 50% could be classified as clinical in nature (ERM, 1994b).

Hospital waste is currently managed as solid waste by area hospitals and no attempt to segregate waste takes place. However, with segregation taking place at source the hazardous waste can be separately handled. This is advantageous, as about 90% of the waste is food waste which can be used in the composting process. Several studies have been conducted on this problem. Only one hospital in the Colombo region has an incinerator of some sort. This however, is not used all the time, as it is not capable of handling the generated waste quantity. In other hospitals the waste gets added to the general collection and is carried into a dump. Pathology wastes are generally buried. The prevailing situation is not acceptable and efforts are underway by the CEIP programme to arrive at a solution.

The total number of hospitals/clinics in Sri Lanka is 510 with 50,091 beds (Ministry of Health data). By considering the waste generation rate per bed to be 0.36 kg/day, the estimated total hospital waste generation is 18 tons per day or 6600 tons per year. This calculation does not include small generators such as small private medical clinics, medical laboratories and surgeries. The ratio has been supported by a study of 23 hospitals in the Colombo metropolitan area, carried out by the Colombo Environment Improvement Programme. These 23 hospitals with 8510 registered number of beds have 3.07 tons per day waste generation.

The workers who handle the waste do not take specific precautions such as wearing protective gear etc., and are exposed to greater risk. They are exposed to the waste shortly after its generation and also have extensive interaction with the waste. It is important that authorities understand the true nature of the risks they endure. At the final disposal points the scavengers too are exposed to higher risks posed by these wastes (Schmidt and Stouch, 1993). Finally, there is a general risk to the population at large as some of these wastes such as plastics, may enter into the market as recycled products.

RESOURCE RECOVERY

As indicated in the comment at the beginning there are scavenger communities active in the Colombo metropolitan area. The scavenging mechanism is organized neatly, with the activity being done by a very poor segment of the population who basically live by the sides of these waste dumps. Out of these landfills, materials such as coconut shells (to make activated charcoal), plastics, metal and glass are recovered. Most of these recovered metal ends up as scrap metal sales to India.

In some situations this system provides an entering level of employment for new urban dwellers emigrating from rural areas. This is well established in the Colombo metropolitan area. The landfilling of waste up to 1995 was carried out at Wellampitiya. The area surrounding this site had a developed system of recovery of valuables. This site also received waste from hospitals and abattoirs, etc., and this aggravated the problem. The Meetotamulla site, which is considered the largest in Sri Lanka, receives the city garbage. The site is situated close to a school.

In addition to these scavenger families living close to dump sites, there exist a host of small-scale garbage collectors, who travel from house to house, collecting recyclable materials such as paper, glass, metals, and plastic containers. This material is then sold to middle-sized garbage collectors who in turn sell the accumulated goods to interested parties. This is not a well-organized material recovery but an informal scavenger sector. Most of the small-scale operators are not aware of the recycling facilities available to them. This limited knowledge of the opportunities open to them restrict them their sales to only a few locations.

Dr. (Mrs) A. Perera of the University of Colombo who had worked on the city's solid waste problem says: "Most of the scavengers are between 11 and 45 years of age. Some are school children. A scavenger's daily income can varies from Rs. 100-500. Most of these boys support their families from what they earn. The relationship between the scavenger community is considered to be most harmonious with each one looking at specific items and not interfering with others. The scavenger community in Sri Lanka is said to be around 10,000" (Perera, 1996).

Her efforts had also resulted in recycling being supported by the co-operative movement of Sri Lanka. The initial collecting centre for recyclable materials was actually opened well outside Colombo, i.e., at Sigiri-Dambulla area. It was begun in 1995 through the Rangiri-Dambulu cooperative society with 21 branch cooperatives taking part in the programme. The villagers can collect any recyclables such as paper, polythene, glass and sell the collection to the co-operative store in their area. The payments range from Rs. 2100 per kg. The goods are transported to Colombo and sold to recycling companies such as the National Paper Corporation, Ceylon Glass Company and Plastic Recyclers (There are about 20 plastic recyclers today mainly in the Colombo district). The Ministry of Co-operatives has extended its full support to the programme. In one city Homagama, about 10 km from Colombo, a program went into operation where a vehicle went around the area on a designated date, collecting waste material and paying spot cash.

With the current paper recycling programme the revenue of the National Paper Corporation had recorded an upward trend so that its income which stood at Rs. 71 million in 1994 increased to Rs. 101 million by 1996. The company is actively supporting the recycling movement and has a published price list for wastepaper. For grades such as off cuts and computer wastepaper the company pays a higher price. The mixed waste bring the least return from the point of view of the wastepaper supplier and this encourages separation at source. Due to this programme tons of paper which had been thrown away by the public as garbage has now ended being reused in the paper making process. It has been established that approximately 70 tons of paper and 15 tons of plastics are being processed daily for recycling either in Sri Lanka or abroad (ERM, 1997).*

The major reason for the slow growth of the recycling industry in Sri Lanka is the lack of awareness among the general public regarding its existence and poor technical support and equipment availability. Secondly, a well-established system does not yet exist in the country for the collection of used items. The co-operative collection mechanism offers a very good beginning as it is well spread across the country. There should be more governmental support to these activities at policy level to breathe in the necessary strength and give much needed impetus to this sector.

PRESENT DEVELOPMENTS

Urban Development Authority (UDA) has worked and formulated a comprehensive environmental management plan for the Colombo Urban Area (CUA) [Table 4]. The Urban Development Authority (UDA), which is a government agency, is responsible for providing standards on buildings, zoning, controlling pollution and maintaining environmental quality. The plan-as options and an action plan—is presented as a summary in the tables (UDA, 1994).

Colombo Municipal Council has already started the privatisation drive with regard to the collection of garbage and the early results indicate much better performance through this mechanism.

The government has understood the need for the necessity for better solid waste management and the preparation of a National Solid Waste Management Action Plan is underway (MTEWA, 1995). The World Bank has extended its assistance to the Government of Sri Lanka. Although the process has been underway for some time with a considerable sum of

---

*Recently however, a slump in the paper industry is Sri Lanka has resulted in loss of activities in paper recycling.

Table 4. Options for improvements of solid waste management.

Urban Solid Wastes

a. Improve the efficiency of the existing urban solid waste collection systems, in CMC, DMMC, Moratuwa MC, and Kotte MC by the following measures:

    1. Purchase of new equipment for increasing the cost effectiveness of collection

    2. Increase the frequency of collection

    3. Increase the capacity of the system

b. Creation of solid waste handling and transfer stations in areas with centralised solid waste collection. The stations to be equipped with installations for inspection, selection, separation and compaction of the solid wastes

c. Provision of a new centralised landfill site for the existing solid waste management systems with adequate environmental protection measures

d. Reduction of waste volumes by composting of waste with a high content of organic matter, e.g. Market waste and consideration of more effective recycling of valuable materials such as paper, glass and metals

e. Separate collection and disposal/treatment of hospital waste

f. Introduce solid waste management in the LAA, where it is presently inadequately organized or nonexistent, with priority for the most urbanised areas. Systems include collection, transfer stations, and transport to the disposal site.

Industrial Solid Wastes

a. Introduction of cleaner technologies to reduce waste generation at the source, if feasible. These steps would reduce industrial solid waste volumes and result in improved working conditions and reduced production costs.

b. Operate a separate collection system for industrial solid wastes with disposal at the municipal disposal sites, with exception of the hazardous wastes which are collected and treated/disposed of separately.

c. Separate collection and adequate disposal and treatment of hazardous waste including hazardous industrial wastewater sludges

d. Creation of a central organization, public or private, responsible for all solid waste management in CMA.

---

money being spent on the study and evaluation process from around 1990s the actual realization of a sanitary controlled landfill site is still some years away. Sites have been shortlisted, investigated, discarded subsequently on grounds of cost of development. Initially, the National Building Research Organisation (NBRO, 1990) undertook an evaluation of more than 300 sites. The first to be selected was a site at Padukka. Opposition from many groups however, quickly lead to dropping the site from any further consideration in 1993

Next, the site chosen was a site at Galudupita marsh, Welisara (ERM, 1993, 1994a), located approximately 10 km north east of central Colombo. The landfill was planned to have sufficient capacity to dispose of the solid waste arising from the CMA area for several years. The plan included one or more transfer stations, through which a proportion of the collected waste arising was to be processed for cost-effective bulk delivery to the landfill site.

This site was selected after a detailed site selection process involving six sites and was not the best site that was available from technical considerations. Sri Lanka suffers from political considerations overriding technical evaluations and the site has been the eventual choice. The syndrome of 'NIMBY' is quite valid and with the political thinking and wishes the optimal site was not selected. Welisara site was an area permanently inundated, low lying marshland with a high groundwater table and periodic flooding in the rainy season and is not at all suitable from a technical perspective.

The intended development at this site was a sanitary landfill with an additional 100 ton demonstration composting facility. The project faced difficulties as solid waste disposal in Sri Lanka has a poor reputation, due to the pollution caused by the existing landfill sites (dumps) in Colombo. The main difficulty is due to the general lack of understanding about the distinction between a dump and a sanitary landfill. It is also perceived by some as associated with the influx of scavengers and low income populations, who earn an income from sorting through the waste for recyclables or hope for resettlement opportunities as the site develops. These negative images had to be overcome to make any solid waste management programme successful.

As the technical feasibility for this site was low, the development cost was calculated to be excessive with result the process has now shifted to Meepe closer to Hanwella. The EIA studies for this site are currently had been completed. This site also is expected to have a composting facility and is expected to serve for 25 years. The landfill gas generated is expected to be flared or vented passively. The present status in this project is the EIA report has been given approval subjected to conditions by the Central Environmental Authority and its awaiting development (ERM, 1997). The residents of the area and several environmental related NGOs however, are opposing the project.

NGO participation

There are about 700 NGOs in the country involved in environmental-related issues. Some of these NGOs are directly interested in solid waste management. For example one NGO based at Moratuwa is actively marketing a simple piece of equipment an organic waste converter, for households. The function of the unit is to turn kitchen waste into compost. The unit has received publicity with the backing of the Ministry of Environment.

In addition, material recycling work is also being promoted. Promotion of the recycling of polythene, glass and metal is being supported. The NGOs are looking at developing the micro-level activities already taking place. NGOs also play a role in educating the people in proper practices.

SWOT ANALYSIS OF SRI LANKA'S SOLID WASTE MANAGEMENT

a) Strengths

Public are aware of the consequences of environmental mismanagement and today they are a strong pressure group. Today they are more demanding of correct practices than earlier and there is a strong driving force for action. There is also the desire among the decision-makers in arriving at a solution.

Solid waste management is covered by National, Provincial and Local Authority Legal/Regulations, as referred to the 13th Amendment to the Constitution (1987), decentralising power to the provinces, and the Provincial Councils Act, No. 42 of 1987. The Municipal Councils Ordinance of the Legal Enactments of the Democratic Socialist Republic of Sri Lanka 1980 is concerned with solid waste management. Section 66 of the Urban Council Ordanance, No. 61 of 1989 and sections 41 and 93 to 95 of the Pradeshiya Sabha Act, No. 15 of 1987 give powers to the Urban Councils and Pradeshiya Sabhas to collect and dispose of waste and derive any income therefrom. Thus there is a basic legal framework existing in the country though precise standards for normal solid wastes are absent. Many action plans have been formulated and steps identified. The multitude of organisations from both governmental and non-governmental sectors is present to support a SWM. The established administrative structure with the local authorities could be effectively utilised in this regard.

There is wide awareness in recycling and interest in resource recovery and reuse. The country has demonstrated that recycling is viable and is being practised in paper and glass and to a limited extent with plastics. There are companies looking at resource recovery and reuse and an informal sector exists to cater to such demands.

Special characteristics of our wastes are also of help. Lower per capita waste generation compared to the developed countries, absence of difficult or complicated waste material (which otherwise would need size reduction), presence of an informal sector engaged in waste recycling are factors that are beneficial in carrying out a management plan.

The presence of an extensive organisation system involved in local government is an asset in implementing a plan. In addition, in Sri Lanka, the local government has received constitutional recognition and thus with proper attitudes and with necessary drive much could be achieved.

b) Weaknesses

A significant weakness is that though necessary legal enactments are in statute books the enforcement and management is not evident in actual practice. Lack of commitment is a serious problem is carrying out a SWM. This perhaps is a malaise affecting the whole system rather than this specific sector. Government should also realise that turning good intentions into realities need investment, committed action and enforcement without any bias.

The established administrative system also could be too cumbersome in the case of handling the hazardous waste component of the industrial waste category. The decision making process today is too slow and lethargic.

There is a lack of technical support towards a strategy for pursuing a proper SWM. There is severe lack in technical support services. We are too dependent on outside sources for finance and are not looking towards innovative solutions (i.e. CESS funds, taxation on products and/or imports for a specific use etc.)

Some other problems faced at present by local authorities are: Lack of dedicated source of budget for SWM. Budget is mainly spent for wages etc. of laborers and there are insufficient funds for proper management; Inadequate revenue base leading to financial problems. Lack of proper planning and proper concept of SWM system. The plans are technically deficient and thus lead to problems. Deficient institutional management without proper job allocation, division of responsibility, and lack of accountability. Mixing health services and related activities with SWM activities under a common budget head (PSs do not have a Public Health Officer in their cadre). Lack of properly qualified, experienced, and trained municipal personnel. Inadequacy of regulatory measures to control and sustain SWM system, for e.g. levy of service charges, revision, and rationalisation of municipal taxes etc. Lack of technical hardware at low cost to spread resource recovery activities at local level.

Many recycling efforts are however; not scientific with people mostly concerned with only the main activity that they are involved with. The informal sector consists of poor or uneducated people. The majority tends to be woman and children. There is much exposure to occupational risk during these practices because they have neither any awareness about health and hygiene nor are they trained to do their work efficiently and profitable. As such a higher degree of efficiency could not be expected from this method. This needs addressing, as this state of affairs is unacceptable.

Improvement of Communication

A recycling program involves and concerns numerous governmental and non- governmental agencies, the public and industry. For a successful waste recycling program, open communication among them is essential. In Sri Lanka the inter agency and intra-agency communication is often poor and this needs to be addressed. There is also poor linkage between the formal waste management units and the informal waste-recycling sector. The awareness of existence of successful practice schemes should be recorded and publicised.

c) Opportunities

There are many opportunities in resource recovery. The organic fraction of our waste is about 70-80% and this can be exploited profitably. Improved awareness should improve recycling practices and methods. There are many research works done in this area and results are available. What we need to implement are some demonstration projects.

d) Threats

The public when the driving force is for survival in a difficult economic situation are not much responsive towards environmental management programs which involves time, effort and continued commitment.

The economics may not support improvements due to such techniques may have to be purchased from outside at high prices.

The small markets provided internally, may prevent recycling practices on economic grounds.

Lack of will power in many sectors to achieve results via small improvements. The desire is for quantum jumps and leap-frogging—some of these buzzwords will be of little or no use. Continuous improvement along sensible paths is the most prudent option.

There is a general lack of faith among the public on policy makers in carrying out implementation. This will be an inertia force in the initial phase of a proper SWM.

SOLUTIONS AND THE FUTURE

To summarize this scenario it is perhaps apt to look at what the future may hold. The present situation is clearly unacceptable and some form of solution is needed quickly. The problem, which is mainly confined to the capital city and other urban areas, is rapidly spreading to the outskirts as well. The appearance of many a scenic place has deteriorated. The following are few pointers to be understood if the country is to save its splendour;

The need to *act rapidly and convincingly* over *endless discussions* is poorly understood. Small-scale beginnings are important. The country is waiting for expensive systems to come into place.

Many people are demanding clean cities, garbage collection and disposal, but only very few are actually ready to help in arriving at solutions. As generators, every citizen has a role to play in the solution. Demands by the citizens without the duty component being understood, it will be difficult to set in place methods to achieve clean cities, roads, etc.

The understanding of the problem is not with everyone. Awareness among people is very poor and the decision-makers themselves are not really aware of the concepts of integrated solid waste management. Garbage is still viewed as a waste to be disposed of and not as a resource. As such the economic evaluations always work against any significant scheme and no returns are (environmental costs and concepts are understood to an even lesser degree) expected.

The infusion of capital is always considered to be from loans or grants. It must be understood that grant money can be best used elsewhere and the issue of solid waste should be addressed with as much of internal funding as possible. The external funded projects have a habit of appearing and disappearing, thus creating only frustration and delays. In most cases, the work gets repeated while costs escalate daily and the problems get compounded. One should also look at self-financing of these projects instead of waiting for big loans.

REFERENCES

Anon (1991), *Biogas process overcomes traditional barriers*, Water & Wastewater International, 16, 2.

de Alwis AAP (1996), *National Survey on Biogas*, The Energy Forum, Vol 1-3, 1996.

EJF (1993), *Citizens' report on Sri Lanka's Environment and Development*, Sri Lanka Environmental Journalists Forum.

ERM (1993), *Solid waste management component: Colombo Metropolitan Area Environmental Project*, Sri Lanka: Inception report, Environmental Resources Management, August 1993.

ERM (1994a), *Solid waste management component: Colombo Metropolitan Environmental Project, Sri Lanka: Selection and Preliminary environmental review of Welisara Landfill site*, Environmental Resource Management, March 1994.

ERM (1994b), *Activity 6 Report, Survey and Disposal Plan for Hospital Waste*. Solid Waste Management Component: Colombo Metropolitan Area Environmental Project, Sri Lanka.

ERM (1996), *Pre-feasibilty study on Hazardous waste management and disposal for Sri Lanka*, Environmental Resource Management, November 1996.

ERM (1997), *Environmental Impact Assessment for a Proposed Sanitary Landfill Alupotha Division, Salawa Estate*, Environmental Resource Management, November 1997.

Fernando DS and Basnayake BFA (1998), The design and development of a vertical composting bioreactor for urban solid waste management, Procs. of Sri Lanka Association for the Advancement of Science Annual sessions, December 1998.

JCI (1981), Japanese Consulting Institute, *Study of composting feasibility for Sri Lanka*.

MTEWA (1995), *Policy and Action Plan 1995-1996*, Ministry of Transport, Environment and Women's Affairs, Sri Lanka.

NAREPP/IRG (1992), *Citizens' views on environmental issues in Sri Lanka*, Public Campaign on Environment and Development, NAREPP/IRG report 1992.

NBRO (1990), Solid waste management in the Greater Colombo Metropolitan Area: Identification of land for solid waste disposal, Colombo, Sri Lanka.

Perera, A. (1996), *Scavengers—a forgotton tribe*, The Sunday Observer, July 21.

Sathianaman, V. (1981), New shortcut method of making compost (Kasala Menik) by the Dalpadado process, Ministry of Agriculture, Sri Lanka.

Schmidt, P. and Stouch, J. (1993), Hospital waste management and Hazardous waste management, NAREPP/IRG staff report.

Tantrigama G (1995), Eco-tourism and conservation of natural areas in the south east dry zone of Sri Lanka, Procs of the Annual Forestry Symposium, Dec 15-16, Hikkaduwa Amarasekera HS and Banyard SG (Eds.), pp. 217-238.

UDA (1994) *Environmental Management Strategy for Colombo Urban Area.*

White, P.T. (1983), *The fascinating world of trash*, National Geographic, April 1983, pp. 426-447.

Section 4

Country Case Studies of European Continent

17

Introduction to Waste Management in Europe

Velma I. Grover and Vaneeta Kaur Grover
411-981 Main St. West, Hamilton, On L8S 1A8, Canada

INTRODUCTION

Europe is a continent of royalties, which has been the mother of many revolutions such as the industrial and telecommunication revolutions, has been the starting place for communism and has the places from where started colonialism and the world wars. Because of the strategic geographical location and natural beauty—land, forests, bio-diversity, mountains, rivers—Europe is a centre of attraction for tourism. The mass tourism, intensified agriculture, massive industrialization has caused many environmental problems like:

- Ozone depletion
- Climate change
- Loss of bio-diversity
- Acid rains
- Pollution in rivers, seas and oceans
- Forest degradation
- Urban stress
- Waste mis-management
- Chemical risks.

Europe and the CIS countries cover more than one fifth of the land area extending between the Atlantic and Pacific oceans—with Russian Federation occupying more than 60% of this total area and hosts 15% of the world population. Europe has faced catastrophes like that of Chernobyl disaster, shrinking of the Aral Sea, acid rains and contamination of surface and groundwater. Besides these problems, waste mismanagement is another major problem associated with urban and industrial area. According to a UNEP report, the average European produces between 150 and 600 kg of municipal waste per year. The West Europeans produce more waste than their eastern counterparts.

WASTE MANAGEMENT IN EUROPEAN COUNTRIES

Baltic Sea Region

Because of the mismanagement of the waste produced, the Baltic Sea has become a very polluted inland-sea. The municipal waste produced by countries around the Baltic Sea is given in Table 1*:

Table 1. Municipal waste in European countries around the Baltic Sea.

Country	Waste produced (tons per year)
Finland	1500000
Estonia	450000
Latvia	870000
Lithuania	1140000
Poland	11400000
Germany	24000000
Sweden	3000000
St Petersburg	1500000

*Anders Bystrom, Composting for Sustainable Development in a Baltic Sea Region Perspective.

Estonia

The waste in Estonia is mainly dumped on the landfill sites—this is said to be influenced by the historical features of waste generation (Juri Teder, Landfills and Landfilling in Estonia). Earlier the main activities were just agriculture, fishing and handicraft. Most of the waste produced from these activities was biodegradable and therefore, was handled either by the people themselves or nature took care of it. But the industrialization, urbanization, large-scale agricultural production and oil-shale mining increased the amount of waste, which included the hazardous waste as well. According to Reigo Lehtla (Waste Legislation Approximation to EU Directives in Estonia) in 1995, 553400 tons of municipal solid waste was generated—that is 335 kg per capita. Almost all of the waste is dumped in an uncontrolled fashion at the unmanaged landfill sites without any separation. There is no separation, incineration or composting. After the Packaging Act there has been a beginning of separation and recycling of paper and glass—but it is at the very initial stages. A lot of waste is exported to other countries, mainly due to the lack of the treatment methods and disposal facilities. For example, in 1995, 264000 tons of metal waste (including 2094 tons of exhausted lead batteries), 5065 tons of waste paper, 238 tons of plastics and 47100 tons of oil-shale fly ash were exported. Some amendments to the Waste Act have been made in 1997 and some new acts have been introduced to manage the waste in a better fashion.

Lithuania

According to statistical data (Irena Gaveniene, Waste Management Policy in Lithuania), 6 million tons of waste was generated in Lithuania in 1996 and it was dumped in an unplanned fashion. There have been plans to have an integrated management system for the hazardous waste and the medical waste but they are still in the planning stages!!!! Most of the waste is sent to landfills and at times it also includes waste from the industries.

Since 1994, at the initiative of Copenhagen Environmental Agency, the recyclable materials (like plastic, paper, metal and glass) are separated at source in different coloured containers. Scavengers also remove recyclable materials from the waste stream.

Latvia

In Latvia, all the waste is dumped at the landfill sites which includes wastes from the industries. In most of the cases an old quarry, wetland or lot of wasteland is used as a landfill site without any kind of environmental assessment. According to Ilze Donina, (Landfill problems and solutions in Latvia), there are around 558 official landfill sites and a few unofficial landfill sites but 90% of the waste collected is disposed of at 48 landfill sites, rest of the sites get only 10% of the waste. At a few landfill sites, compaction is carried out on a daily basis and waste is covered with soil, at others if carried at all it is done on a monthly basis. There is no other management of the landfill sites and the scavengers collect the recyclable materials from the landfill sites.

There are plans and efforts to make solid waste management plans, secure landfill sites, separating hazardous waste and to pass some Acts— to make solid waste management effective.

Sweden

The municipal waste in Sweden is divided into household waste, construction and non-sector-specific waste, garden waste, construction and demolition waste, sewage sludge, and waste from vehicle scrapping. The Table 2 (tabulated from the data given by Karin Oberg, Swedish Environmental Protection Agency, Sweden Action Plan), gives the amount of waste dumped on landfills, amount incinerated, recycled, and biological treatment.

Table 2. Waste Management in Sweden.

Type of Waste	Landfilled	Recycled	Incineration	Biological treatment
Household waste	39%	16%	42%	3%
Construction and demolition waste	75%	–	–	–
Non-sector specific waste	70%	–	–	–
Sewage sludge	65%	–	–	–
Garden waste	50%	–	–	50%

Of the 300 landfill sites in the country, more than half of them have the leachate recovery system and gas recovery systems have been increasingly installed at number of these sites. Because of the stricter Acts, recycling has increased in the last few years—the main recycling is of the packaging material but there has been some collection and reclamation of metals also. Composting is wide spread in the houses with large gardens but municipalities also have composting plants and biological gasification plants (digesters).

Most of the biological waste and the sewage sludge are sent to these plants. Incineration is used to generate energy, which is utilized for heat production and is delivered to the local district heating network or nearby industries. As quoted by Karin Oberg, the quantity of the waste incinerated has doubled between 1980 and 1994 and energy production has more than trebled.

Spain

The data from the Environment Ministry of Spain shows that in 1994, 14300000 tons of solid waste was produced—giving yearly per capita generation as 994 kg. According to Jorge Gallego Rubio (The Role of Landfills in the Current Management of Household and Industrial Waste in Spain), almost 59% of the household waste is dumped at controlled landfills, 12% in treated compost plants and around 4% is incinerated. The rest of the waste—around 3.5 million tons, is dumped in an uncontrolled manner. It is quite obvious that landfill is the most common disposal method. The separation of waste at source was started some years ago—but its impact is still fairly insignificant. The main problem that needs to be addressed in Spain at the moment seems to be to control the uncontrolled dumping, prevent industrial waste from going to landfills meant for household waste and improve the conditions of existing controlled landfills.

Hungary

According to Zsuzsanna P Koltai (Environmental Issues Concerning Waste Management in Hungary), around 4 million tons of the solid waste is generated annually in the country. Approximately, 82% of this waste is collected and disposed of in the following manner:

Landfill	75%
Incinerators	8%
Recycling	3%
Unknown	14%

There are around 2682 landfill sites, out of this 900 work without official permits. Of those who operate with permits, only 100 of them comply with the environmental regulations.

England

The total waste generated in England is around 435 million tons per year. Out of this only 5% is household waste, 8% sewage sludge, and 16% construction waste. The household waste is approximately 20 million tons which is around 0.45 tons per head. At present 90% of the household waste is dumped on the landfill sites, 5% is incinerated and 5% is recycled. The policy stresses on the recovery of energy from waste—but incinerators are being discouraged.

Germany

Germany is one of the first countries which paid attention to the waste management problems and has come a long way since 1972. In 1972 waste disposal was governed for the first time by uniform national regulations through the implementation of the Waste Management Act (Abfallwirtschaftsgesetz). This defined binding standards for the collection, treatment and disposal of various types of waste and has since been amended several times in accordance with best available technology. In 1986 an order of priorities for dealing with wastes was legally defined: avoidance, recycling, disposal.

At the outset of the 1990s the concept of closed substance cycles was successfully used to reverse the trend towards growing waste volumes. In Germany there was some 10% less waste in 1993 (337 million tons) compared with 1990. The total volume of waste in that year included 143 million tons of waste building materials, 78 million tons of production wastes, 68 million tons of mining material, 43 million tons of domestic refuse and 6 million tons of other wastes. The recycling quota also improved in the period 1990/1993 from 20 to 25%.

After the new Closed Substance Cycle and Waste Management Act (Kreislaufwirtschafts- und Abfallgesetz), which came into force in 1996, only wastes which cannot be recycled or recovered may be disposed of and this must be done in a sound way. From 1999 enterprises above a certain size will be obliged to submit waste life-cycle analyses.

The 1991 Packaging Ordinance (Verpackungsverordnung) is a prototype for legislation designed to close substance cycles. Because of this Act, in the period from 1991 to 1995 the use of sales packaging in private households and small-scale industry declined by 12% (from 7.6 to 6.7 million tons). This should be seen in the context of the steady growth in the use of packaging observed from 1988 to 1991. In 1995 each citizen, with a per capita consumption of 82 kg, was using on average 13 kg less packaging than in 1991.

Moreover, a high level of material recycling was achieved for a great deal of the sales packaging (figures for 1995):

- 87% for paper and board
- 78% for glass,
- 67% for aluminium
- 61% for tin plate,
- 58% for plastics and
- 49% for composites.

In addition, the Packaging Ordinance protects the systems of returnable beverage packaging, which are traditionally well developed in Germany, by stipulating a 72% share for returnables in the beverage sector. If this quota is not reached, the Packaging Ordinance requires a compulsory deposit on all one-way beverage packaging.

The principle of cooperation between government and industry has proved particularly valuable in the sphere of waste avoidance. A prime example of such voluntary commitments by industry in the waste sector is the agreement to take back and recycle used batteries, waste paper, building waste and scrap cars. The collection and disposal of wastes, including the planning of waste treatment facilities, comes under the authority of the Länder (States): However, the Federal Government lays down nationally uniform requirements for the disposal of wastes.

In 1993 an annual volume of some 9 to 10 million tons of human settlement-generated wastes were incinerated in 51 domestic refuse incineration plants serving an area with 23.8 million inhabitants. This corresponds to approximately 30% of overall household waste. Despite the decline in waste volumes, the recycling and disposal infrastructure in Germany needs to be further developed in order to establish effective and lasting risk prevention in the waste sector. Moreover, reliable and modern waste treatment is an important economic factor and an important prerequisite for attracting new businesses.

Italy

The problem of waste management has taken a new importance in Italy over the last decade due to the tremendous increase in the quantity of the waste generated and its qualitative evolution. It is estimated that at the domestic level, each citizen produces almost 1 kg of urban solid waste daily, thus for the entire nation the waste production amounts to 20 million tons per year. All of this waste is not collected and disposed of properly— only 40% of the urban waste generated is disposed of in properly managed facilities (Towards new Environmental Solidarities, Azienda Servizi Municipalizzati Brescia). The remaining waste is either abandoned or processed at unsuitable sites.

Yugoslavia

The waste collected from Yugoslavia is mixed and is mostly disposed of on landfill sites. Legislation requires the segregation of waste at source and the recyclable materials are to be separated by the consumers. This has been started but is not yet fully operational. In addition to this composting— anaerobic digestion and co-digestion of the organic and biodegradable waste is being encouraged.

CONCLUSION

Europe produces around 250 million tons of waste per year. The quality of waste generated in some of the European cities can be seen in the Table 3 compiled from various reports:

Despite the increased emphasis on waste prevention and recycling, most European waste is disposed of in landfill and incineration. Figure 1 (see also Table 4) gives the comparison of the waste disposal cost ($/ton) on landfill and incineration plants in various countries.

Table 3. Waste composition in some European cities.

Location	Paper	Plastic	Metal	Rubber	Leather	Textiles	Glass	Organic	Miscellaneous
Belgrade, Yugoslavia	16.56	7.61	4.93	1.75	1.59	1.71	9.15	36.73	19.97
UK	33	7	8	–	–	4	10	30	8
Rome, Italy	18	4	3	–	–	–	4	50	21
Lithuania	13.4	6.78	4.18	–	1.92	5.4	10.17	46.6	11.32
Spain	21.6	10.57	4.12	1.02	–	4.82	6.88	44.09	7.02

Table 4. Comparison of MSW incineration in Europe (1991 data).

European countries	Amount of waste incinerated (Mg × 10³ per year)	Incinerated (%)	Incinerated with energy recovery (%)	No. MSW incinerators (%)
Austria	300	18	18	2
Belgium	720	1	7	27
Denmark	1500	70	70	48
Finland	Np	Np	Np	1
France	6350	40	13	260
Germany*	9300	23	22	49
Greece	15	Np	Np	1
Ireland	—	—	—	0
Italy	2000	10	4	54
Luxembourg	140	78	78	1
Netherlands	2805	46	36	11
Norway	440	23	13	50
Portugal	—	—	—	0
Spain	697	6	43	22
Sweden	1550	55	55	22
Switzerland	2300	80	60	49
United Kingdom	2780	10	3	31
Total				629

* Original and new federal states.
n.p. = not published.
Source: Johnke (1992)

Table 5. Future legislation for municipal solid waste disposal.

Country	Legislative Objectives
Austria	Aims to ban landfill of organics (>5%) by 2004
Switzerland	Banning landfill of combustible waste
France	Banning landfill of combustible waste by 2000
Denmark	Banning landfill of combustible waste by 1997
The Netherlands	Landfill of combustibles, paper and cardboard banned. Landfill of municipal solid waste banned by 2000
UK	Recycling target of 25% by 2000. Currently achieves 6%.

Source: Whiting Kevin J and F J Schwager (European Trends in the Thermal Treatment of Solid Wastes) in The ISWA Year Book 1997/1998.

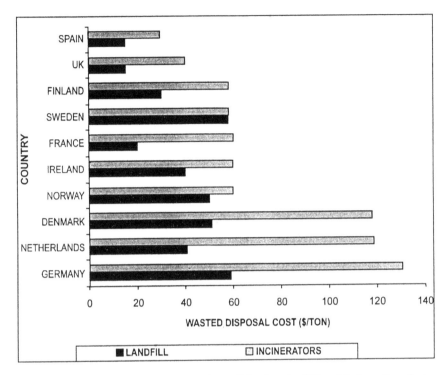

Fig. 1. Comparison of the waste disposal cost ($/ton) on landfill and incineration plants in various countries
Source: Europe's Environment: The Dobris Assessment—An Overview (Eds—David Stanners and Philippe Bourdeau)

Table 4 gives the comparison of MSW Incineration in Europe (1991 data)

With a lot of public protest and pressure from non-governmental organizations many incinerators have been closed down and are becoming increasingly unpopular. The more recent laws lay emphasis on recycling and reuse. Table 5 gives the future legislation for municipal solid waste disposal in some of the European countries are:

From the above discussion it can be seen that the western European countries have more legislation for the waste collection and segregation. The landfill sites are better engineered and there has been more recycling but in the eastern European countries also, now there has been an increased effort to manage the waste.

REFERENCES

Anders Bystrom, Composting for Sustainable Development in a Baltic Sea Region Perspective.
Europe's Environment: The Dobris Assessment—An Overview (Eds—David Stanners and Philippe Bourdeau)

Europe's Environment: The Dobris Assessment—An Overview (Eds—David Stanners and Philippe Bourdeau)

Johnke (1992).

Jorge Gallego Rubio, The Role of Landfills in the Current Management of Household and Industrial Waste in Spain

Karin Oberg, Swedish Environmental Protection Agency, Sweden Action Plan

Towards New Environmental Solidarities, Azienda Servizi Municipalizzati Brescia.

Whiting Kevin J. and F.J. Schwager (European Trends in the Thermal Treatment of Solid Wastes) in The ISWA Year Book 1997/1998.

Zsuzsanna P. Koltai, Environmental Issues Concerning Waste Management in Hungary

18

Waste to Energy—The UK Situation

Stuart G. McRae

Wye College: University of London, Ashford, Kent TN25 5AH, UK

INTRODUCTION

The total amount of waste generated annually in the UK is about 435 million tons (Department of the Environment, 1995b), but a large proportion of this is agricultural waste (mainly by-products from housed livestock) or wastes from mining and quarrying (Fig. 1). The rest (245 million tons) is referred to as "Controlled Waste" whose treatment and disposal is covered

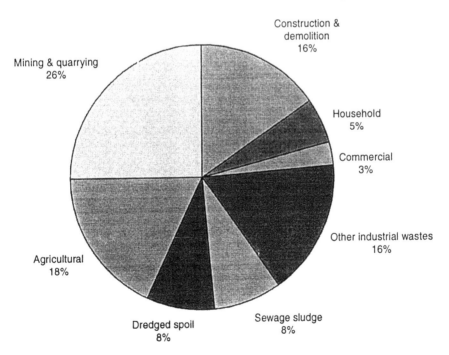

Fig. 1. Types and amounts of waste generated annually in the UK

by specific legislation, most notably the Environmental Protection Act of 1990.

Household or municipal waste (20 million tons in total, or about 0.45 tons per head of population annually) refers to the waste generated by domestic properties. Most of it is collected, usually weekly, by a service funded by local taxes. The 20 million tons includes about 5 million tons of waste, usually large, bulky items which the public take to designated locations called civic amenity sites. Most household waste in the UK consists of paper, packaging and vegetable matter including food wastes (Table 1).

Table 1. Typical composition of UK household waste (in per cent).

Paper and board	33
Plastic	7
Textiles	4
Metals	8
Glass	10
Vegetable matter	30
Miscellaneous	8
	100

Commercial waste, collected from shops, offices, schools, etc., generally has a similar composition. Industrial waste is of various kinds including, for example, construction and demolition wastes, pulverized fuel ash from power stations, blast furnace slag, and worn-out motor vehicle tyres.

The collection and disposal of controlled waste is now undertaken entirely by private companies, although until a few years ago household waste was directly collected and disposed of by local government agencies. The local government is now only responsible for arranging for the collection of household waste by private companies, for preparing an overall waste management strategy for their area, including making provision for waste disposal facilities, and for licensing and inspecting waste management and disposal facilities such as landfill sites, incinerators and waste transfer stations where wastes are locally concentrated for bulk transport or treatment.

THE UK WASTE MANAGEMENT STRATEGY—REALITIES AND OBJECTIVES

Currently about 90% of household waste is disposed of to landfills, 5% is incinerated and 5% is recycled (Department of the Environment, 1995b). A slightly smaller proportion of commercial waste is landfilled (about 85%) with correspondingly more incinerated (7.5%) or recycled/reused (7.5%).

The reason for the dominance of landfilling as a waste disposal option in the UK has been and still is principally its low cost compared with other methods of disposal. This is due mainly to the plentiful supply of suitable void space left by mineral extraction in most parts of the country, though there are problems in some areas. Operating costs are also relatively low. During the 1980s, for example, typical costs of disposal by landfilling were four or five times less than disposal by incineration. Landfilling prices have recently increased considerably because of increasingly strict environmental standards. Most modern landfills are required to be fully engineered containment sites with natural or synthetic liners and capping layers, and fitted with full leachate and gas monitoring and control equipment (Department of the Environment, 1995a).

Typical costs of landfill disposal are now (1996) around £ 15-20 per ton of delivered waste. The introduction of a landfill tax of £ 7 per ton in October 1996 will also increase the costs. Nevertheless, it is predicted that the gap between the cost of landfill and incineration, typically around £ 30 per ton of delivered waste in a modern unit with energy recovery, is unlikely to close any further in the near future so that disposal by landfilling will remain the cheapest option. The relative advantages and disadvantages of landfilling are set out in Table 2.

Table 2. Advantages and disadvantages of landfilling

Advantages	Disadvantages
Relatively cheap	Becoming more expensive; can be very expensive in the long term if it causes water contamination or landfill gas emission problems
Large capacity available in some areas, and waste management industry very "geared" to landfill disposal	Some areas very short of suitable sites and wastes have to be transported long distances
Suitable for a wide range of wastes	Versatility and convenience make it less attractive for waste producers to be innovative in the way they deal with wastes
Restored land can be used beneficially	Restored land can be contaminated or unsuitable for some after-uses
Can be unobtrusive if well designed	Landfills may be a source of noise, dust, odours and vermin
Can be a source of landfill gas which can be used as an energy source	Energy recovery from landfill gas is not very efficient
	Landfill gas can be a hazard and is a harmful "greenhouse gas" if vented to the atmosphere

Recycling waste in order to process it into usable raw materials or products is, and is likely to remain, a relatively small scale method of dealing with wastes, though there are some exceptions. For example, in some committed local communities over 80% of municipal waste is recycled. The economics of recycling are highly volatile, and almost all recycling schemes are kept viable by a subsidy, called recycling credits, essentially a reward for removing material from the general waste stream which would otherwise go mainly to landfill.

The current practice in the UK, therefore is virtually the opposite of the Government's preferred strategy which is based on a Waste Hierarchy (Fig. 2) in which methods of dealing with wastes are arranged according to how well they conform to an overall Sustainable Development Strategy (Department of the Environment, 1995b). The first priority for a more sustainable waste management is declared to be waste REDUCTION, followed by REUSE (e.g. retreading tyres or having reusable rather than throw-away bottles) and RECOVERY, a broad strategy incorporating recycling, composting organic wastes and energy recovery from waste. At the bottom of the hierarchy is waste DISPOSAL, essentially by landfilling (but including incineration without energy recovery), regarded as the least environmentally sustainable, although currently by far the most common method of dealing with the U.K.'s waste.

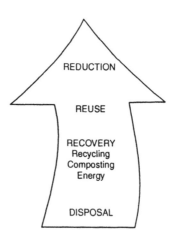

Fig. 2. The U.K.'s waste hierarchy strategy.

Economic considerations are taken into account in the concept of the Best Practical Environmental Option (BPEO) which is the option that "provides the most benefits or least damage to the environment as a whole, at acceptable cost, in the long term as well as the short term". This takes into account both the environmental and the economic costs and benefits of different options. For example, in some circumstances the BPEO might actually be landfill where it results in the effective restoration of mineral workings to a beneficial after-use.

A further factor which has significantly affected the economics of waste to energy projects in the UK in recent years is the Non-Fossil Fuel Obligation (NFFO). This is a Government requirement for each electricity company to derive a specified but relatively small proportion of its electricity from non-fossil fuel sources (wind, hydro, landfill gas, municipal and industrial waste, energy crops and agricultural and forestry wastes). A premium price is paid for electricity generated in this way.

INCINERATION IN THE UK

Incineration of municipal waste was pioneered in the UK, in the late 19th century, initially as a method simply for dealing with wastes in urban areas but subsequently, to a limited extent, as a means of electricity generation. This function became much less important with the development of the National Grid for the distribution of electricity from large coal-burning power stations.

In 1991, it was estimated that there were almost 800 waste incinerators in use in the UK, but by far the majority of these (700) were situated in hospitals for the treatment of clinical wastes. Most of the others dealt specially with wastes produced within particular industries, including four high-temperature incinerators dealing with certain kinds of hazardous wastes. There were 30 which burned municipal wastes. By 1994, this number had declined to 24, and was expected to fall still further to six or seven by the end of 1996. The main reason for this decline was the inability of many of the incinerators to meet the increasingly stringent emission limits required by European Community Directives. The various perceived advantages and disadvantages of incineration are set out in Table 3.

In spite of the many disadvantages cited in Table 3, one "new generation" incinerator has been built and is in operation and several others are under construction or being planned. All of these have energy recovery capability and this is likely to be the norm for any future incinerators so as to partly offset the high capital and operating costs. The Energy from Waste Association argues that another 10 to 15 plants may be needed by the end of the century.

One such "new generation" incinerator is the South East London Combined Heat and Power (SELCHP) facility (ETSU, undated). This is situated about 6 miles from central London and is designed to deal with the municipal wastes from about 400,000 homes and commercial wastes from local businesses (Fig. 3). It uses well-established mass-burn technology

Table 3. Advantages and disadvantages of incineration (including waste to energy).

Advantages	Disadvantages
Can be built close to the source of waste, reducing transport costs	High capital and operating costs makes it relatively expensive
Suitable for many flammable, volatile, toxic and infectious wastes which should not be landfilled	Reliance on incineration could restrict choice of future disposal options, including a proper consideration of waste minimization or recycling
Renewable source of energy, yielding up to five times as much energy per ton of waste than energy recovery from landfill	Poor energy efficiency as compared with fossil-fuels; raw materials and energy are consumed to replace the incinerated items; unlikely ever to make significant contribution to the UK energy budget
Produces no methane (unlike landfills)	Significant danger of atmospheric pollution (though modern generation incinerators do meet strict emission standards); some incinerators generate toxic liquid effluent
Reduces volume of waste requiring disposal to landfill	Volume of residue still about 40-50% of equivalent waste in compacted landfill
Materials recovery is possible from the unburned residue	Concentrates toxic materials in the residues
Minimizes the landfilling of wastes (regarded as environmentally unfriendly)	Most residue still has to be landfilled
	Incineration is deeply unpopular with communities close to proposals

for the combustion of 420,000 tons of waste per annum and generates 30 MW electricity supplied to the National Grid with future plans to sell excess heat to 7,500 local homes via a District Heating Scheme.

Waste is delivered to an enclosed tipping hall capable of storing up to four days supply (to cover weekends and holidays), and is then fed by a crane into one of the two identical incinerator streams each capable of burning 29 tons of waste per hour. The furnaces use reverse acting stoker grates to agitate the waste during combustion. Energy recovery is by means of boilers and economizers producing superheated steam which drives a single turbo-alternator generator. To minimize pollution, the temperature of combustion is carefully controlled and the plant uses flue-gas cleaning equipment including acid-gas scrubbers and bag-filters which remove particulates and dust. The total cost of the SELCHP plant was about £ 100 million. Operating costs are estimated at about £ 7 million per year including labor, consumables and residue disposal costs. The electricity produced is supplied to the Regional Electricity Company under the provisions of the NFFO.

However, public reaction to proposed new incinerators is generally adverse and results in a classic NIMBY (Not In My Back Yard) response.

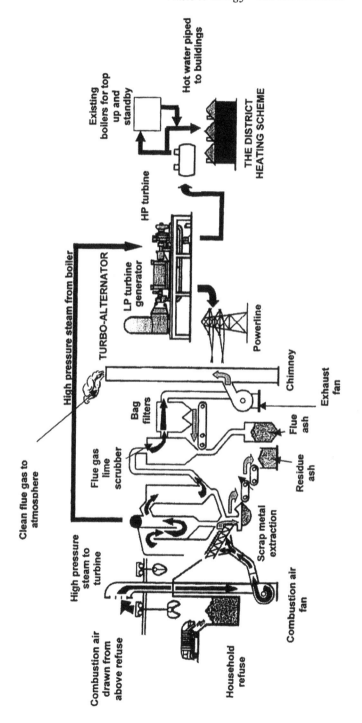

Fig. 3. Diagram of the South East London Combined Heat and Power Plant

While the public expect their wastes to be removed regularly, they often are unaware of what then happens to them and are not really too concerned about the methods of dealing with them unless a waste treatment or disposal facility is proposed near their homes. Opposition to incinerators has been particularly strong, mainly on the grounds of potential atmospheric pollution, and that building more incinerators simply postpones the day when more environmentally friendly and morally justifiable options such as waste minimization, recycling or reuse are employed (Collis, 1996).

For example, an active and vociferous protest group is opposing a suggestion to build the first of three incinerators to burn household and other wastes in the county of Kent, SE of London. Some of their propaganda, headed "Cancer, Birth Defects, Breathing Problems, Infertility" claims that they are about to be "surrounded by poisonous emissions, 24 hours a day, 365 days a year, for the next 25-30 years" and that these include a "soup of chemicals, many of which are extremely dangerous if breathed in or eaten", especially by those members of society "least able to defend themselves including developing foetuses, young children and senior citizens". Irrespective of whether or not these claims are true, the campaigners putting them forward do believe them and are seeking to persuade others to adopt the same point of view.

An alternative to mass-burn incineration is the production of pelletized "refuse derived fuel" which can be burned in conventional furnaces for heating or drying processes as a total or partial substitute for coal. However, there are problems with boiler corrosion, and this process is not widespread in the UK. Recently, the possibility of using household or commercial wastes as a fuel in some industrial processes such as cement manufacture has been investigated. If adopted, they could be the BPEO for some waste streams and the UK would then be following more common practice in the US and the rest of Europe.

ENERGY FROM LANDFILL GAS

The decay of organic matter within landfills produces gas. In a mature landfill the gas is approximately 30% carbon dioxide, 55-60% methane and 10-15% nitrogen together with many other trace components, some having unpleasant odours. This gas is an explosion hazard if it accumulates in confined spaces, for example in buildings on or adjacent to the landfill site, and it is also harmful to vegetation. Thus, it has to be controlled and all modern UK landfills containing putrescible wastes are fitted with gas monitoring and control equipment, together with monitoring and control equipment for leachates.

In some cases, the landfill gas is allowed to vent directly to the atmosphere but this is discouraged since methane is a significant "greenhouse gas", and venting to the atmosphere does nothing to control odours. More commonly, however, the gas is collected by a series of gas wells (Fig. 4) and piped to a location where it is simply flared off (Fig. 5) or where it is used as an energy source. (Cooper *et al.* 1993a, 1993b; Department of the Environment 1991; National Association of Waste Disposal Contractors 1994).

In early systems, the gas was directly burned in industrial processes close to the landfill, for example, as a fuel for brick-making. The Non-Fossil Fuel Obligation (mentioned above in the context of incineration), however, has provided an incentive to use landfill gas for electricity generation. This has also meant that landfills away from any direct user could be tapped as an energy source. Gas turbines, dual fuel (compression ignition) engines and spark ignition engines are all used, ranging in size from a few hundred kW to several MW. Energy recovery from landfill is considerably less efficient per ton of waste than incineration but landfills are often in less

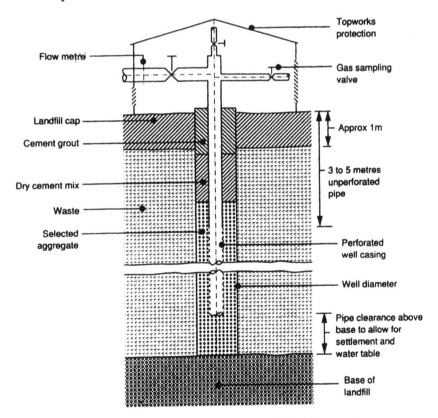

Fig. 4. Cross-sectional drawing of a typical landfill gas well.

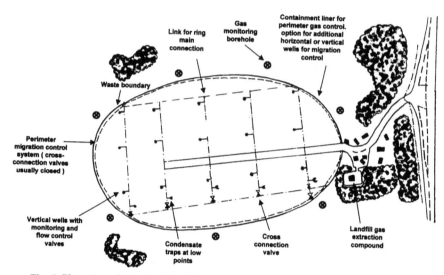

Fig. 5. Plan view of a typical landfill gas collection system with flare compound (an electricity generation plant can be substituted).

sensitive areas so that there is less concern over possible atmospheric emissions. The gas yield depends on the amount of waste, its composition and site-specific factors, principally the water regime. A large modern landfill will probably produce usable gas for over 15 years.

ENERGY FROM SEWAGE SLUDGE

The UK has an extensive system for collecting and treating human wastes. This annually produces about 35 million tons (wet weight) of sludge, with about 4% solid matter. At present, about half is recycled onto agricultural land, 30% is disposed of to the sea (a method being rapidly phased out) and the rest is incinerated or landfilled. Most sewage undergoes an anaerobic digestion process (Baldwin, 1993) and energy can be recovered by utilizing the methane produced (Frost, 1993). There are about 120 combined heat and power plants which run on methane from sewage, with the electricity used to power the works or sold to the National Grid. However, many plants are very inefficient and produce little gas for energy purposes.

ENERGY FROM AGRICULTURAL AND FORESTRY WASTES

These wastes, including slurry and wastes from housed livestock and bark and chippings from the forestry industry, fall outside the UK's legal

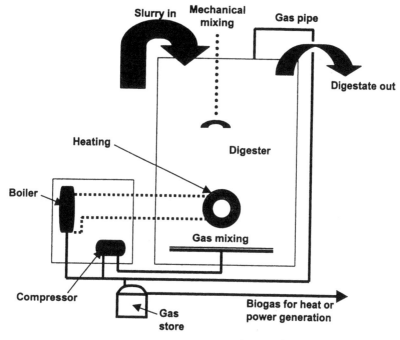

Fig. 6. Diagram of a typical farm-scale anaerobic digester

definition of controlled waste and are regarded as industrial by-products. Most of these wastes are eventually disposed of to the land as fertilizer or soil conditioner, but some are used as fuels in waste to energy systems. Examples include small-scale, on-farm anaerobic digesters for farm slurries, whose main purpose is for pollution control, but which also produce biogas which can be used to power small electricity generators (Fig. 6). There are also a few larger-scale facilities collecting waste from a number of units, for example a recently constructed power station fired by poultry litter.

THE FUTURE

In addition to its general Waste Strategy which tends to encourage incineration with energy recovery rather than landfilling (Department of Trade and Industry, undated; ETSU, 1996), the UK government is actively supporting renewable energy, including energy from waste through the New and Renewable Energy Programme of the Department of Trade (Department of Trade and Industry, 1994). However, at the moment renewables provide less than 1% of the primary energy requirements of the

UK and less than 2% of electricity supply. Hydro-electric power is responsible for about two-thirds of the latter. Although NFFO is seen as a major incentive for an increased amount of energy from waste in the future, this should be viewed more as a mechanism for improving the economics of waste treatment options rather than encouraging waste to energy as a major source of energy.

Nevertheless, waste to energy, particularly incineration, still has to overcome the major disadvantages of higher costs and adverse public perception. Thus, the likelihood is there will be some increase in waste to energy schemes in the future, including increased utilization of landfill gas, but there will be no dramatic or sudden change in the ways of dealing with the UK's waste. Much, however, depends upon political will and public opinion, both rather uncertain commodities.

REFERENCES

Baldwin, D.J. (1993). *Anaerobic Digestion in the UK—a review of current practice.* Report B/FW/ 00239/REP. ETSU, Harwell, Oxfordshire OX11 0RA, UK.

Collis, C. (1996). *Pyromania.* The Ethical Consumer, June 1996, 22-25.

Cooper, G., Gregory, R., Manley, B.J.W. and Naylor, E. (1993a). *Guidelines for the Safe Control and Utilization of Landfill Gas. Part 4A. A brief guide to utilizing landfill gas.* Report B 1296-4A. ETSU, Harwell, Oxfordshire OX11 0RA, UK.

Cooper, G., Gregory, R., Manley, B.J.W. and Naylor, E. (1993b). *Guidelines for the Safe Control and Utilization of Landfill Gas. Part 4B. Utilizing landfill gas.* Report B 1296-4B. ETSU, Harwell, Oxfordshire OX11 0RA, UK.

Department of the Environment. (1991). *Landfill Gas: A technical memorandum providing guidance on the monitoring and control of landfill gas.* Waste Management Paper 27 (Second Edition). HMSO, London.

Department of the Environment. (1995a). *Landfill Design, Construction and Operational Practice.* Waste Management Paper 26B. HMSO, London.

Department of the Environment. (1995b). *Making Waste Work. A strategy for sustainable waste management in England and Wales.* Cm 3040. HMSO, London.

Department of Trade and Industry. (1994). *New and Renewable Energy: Future prospects in UK.* Energy Paper No. 62. HMSO, London.

Department of Trade and Industry. (undated). *The DTI Energy from Waste Programme.* (leaflet) ETSU, Harwell, Oxfordshire OX11 0RA, UK.

ETSU (for the Department of Trade and Industry. (1996). *Energy from Waste: a Guide for Local Authorities and Private Sector Developers of Municipal Solid Waste Combustion and Related Projects.* Revised Edition. ETSU, Harwell, Oxfordshire OX11 0RA, UK.

ETSU (undated). *Energy from Municipal Solid Waste: SELCHP, Lewisham.* (leaflet ECS2). ETSU, Harwell, Oxfordshire OX11 0RA, UK.

Frost, R.C. (1993). *Energy Recovery from Sewage Sludge in the UK. Current situation, prospects and constraints.* Report B/MS/00192/REP12. ETSU, Harwell, Oxfordshire OX11 0RA, UK.

National Association of Waste Disposal Contractors. (1994). *Guidelines for Control and Utilization of Landfill Gas.* NAWDC Landfill Guidelines 15. NAWDC (now ESA), Mountbarrow House, 6-20 Elizabeth Street, London SW1W 9RB, UK.

19

Environmentally Compatible Strategies for the Integrated Management of Municipal Solid Wastes

Joan Mata-Alvarez

Department d'Enginyeria Química i Metal·lúrgia, Facultat de Química
Universitat de Barcelona, Martí i Franquès 1, 08028 Barcelona, Spain

INTRODUCTION

Man has constantly sought to improve the quality of life. However, while technology and innovation have played an obviously vital role in achieving this goal, the increasing generation of wastes has been the price of this improvement. Today it is estimated that more than 1.2 kg of municipal solid wastes (MSW) is generated per day per person in many western countries, and despite the fact that in recent years MSW-disposal issues have come to the fore, this rate is still climbing.

Nevertheless, there is an evident change in mindset as people realize that this wasteful society has to come up with solutions to manage its resources and its residues. Thus, more stringent environmental regulations favouring the design and managment of landfills and incinerators have been implemented in most developed countries. Concepts such as integrated waste management, which means the combination of the best available technologies to achieve the proper management of the wastes, have been introduced. As a consequence of these issues, many programmes aiming at the collection, processing and sale of recyclable materials have started. Among other reasons, recycling of MSW is attractive to legislators because of the benefits arising from the energy saved by reusing certain raw materials found in MSW and from reduction in landfill volume used. Nevertheless, not only recycling is important; checking the formation of MSW right at the origin is obviously the most desirable solution, yet, over the years, it seems to have received little attention. Minimizing residues through reduction at source and recycling are topics of importance for environmental technologies. There is wide scope for applying such principles. In the specific case of MSW the reduction at source must be based, on the one hand, on lower consumption, and on the other, on modifying the life cycle of products. Recycling of MSW not only implies that of the components (paper, plastics, glass, etc.), but also that of the goods themselves. Following the hierarchy of technologies, the next that follows is waste-to-energy conversion and, finally, landfilling, when no other further use of the waste can be envisaged.

The organic fraction of MSW (OFMSW), and more specifically, the biodegradable fraction, has to be recycled through composting. Through such a process, stabilized organic matter can be returned to the soil with all its macro- and micronutrients. Composting is a practice with a long history in Europe. Compost commercialization, however, is difficult, as quality is usually neither high nor constant and there are other alternative options which are cheaper.

In subsequent sections, ways in which society can change the manner in which it handles its wastes, is discussed. This includes possibilities of more efficient use of materials and better design procedures. Both are options that direct MSW away from landfill disposal and waste-to-energy procedures. The first part deals with reduction and recycling strategies for municipal solid wastes. The second part deals with technical innovations which have made comprehensive strategies for dealing with the area of organic fraction of MSW management.

STRATEGIES FOR SOURCE REDUCTION OF MSW

The best option to minimize the volume of MSW is to reduce the production, of the wastes right at the origin. Source reduction policy should start with general public education. The investment on spreading the knowledge of the links between resource conservation and waste production should provide benefits as it is more likely that well informed people will follow any MSW minimization programme. Some possible measures to improve the level of public awareness towards reduction of municipal solid wastes are presented in Tables 1 and 2. Especially significant is the potential for the reduction of packaging pointed out in Table 1. For instance, in the USA, approximately 30% of the MSW consists of packaging material (Sikora, 1990). Paper and paperboard make up for nearly half of this percentage. It follows that there is a large potential for manufacturers to reduce this quantity, for example, by trimming the weight of the packaging materials. In addition, these measures involve a reduction of costs for companies. In accordance with EPA, industries can reduce around 10% of the raw materials they use, by improving the process efficiency and by reducing the package (Anderson and Burnham, 1992).

Table 1. Issues for source MSW reduction.

Educational	Industrial
Individual face-to-face information	Manufacture of goods more repairable
Primary School Education: Development of specific topics	Use of less packaging. Concentration of products.
Public mass-media campaigns	Reuse of packages
Training public waste managers in new technologies	Use of recycled materials
	Inclusion of environmental information on packages

Table 2. Some political issues for MSW source reduction.

Changing taxing systems (Ex.: Higher taxes for products difficult to recycle)
Resource management (Ex.: Introduction of a packaging tax and in general, taxing energy and resources)
Funding research for waste reduction
Construction/design and production (see Section "A sustainable approach for MSW management")
Services (Ex.: Tax exemption for repair)
Transportation/distribution (see Section "A sustainable approach for MSW management")
Consumption (Ex.: Development of ecologically oriented shopping)
Disposal (Ex.: Adoption of tax collection incentives: The more you dispose, the more you pay)

In examining the issues presented in Tables 1 and 2, it is interesting to consider the study carried out by Gilnreiner (1994). The author analysed the public acceptance of measures regarding minimization taking into account social and ethological aspects. He concluded that the reduction of MSW can only be achieved if the consumer is provided with measures that make it easier to consume less and change his consumption habits. This is a difficult task as it is relatively easy to trigger the consumers appetite. Human behaviour seems to support economic growth and is antagonistic to waste minimization. In accordance with this author, measures like those presented in Tables 1 and 2 are difficult to accept. Especially those regarding imposition through public relations and laws, which, in accordance with this author, are likely to result in dissatisfaction and failure. The analysis carried out by Gilnreiner (1994) for future trends, taking as a basis the data from the City of Vienna, posed a limit of approximately 10% of household waste fraction that can be reduced due to waste minimization.

STRATEGIES FOR RECYCLING MSW

The second alternative in the hierarchy for MSW management is recycling. Recycling involves several steps which can include the separate collection and/or the sorting processes, the transport, the pre-processing (compaction) and the recovery process. Table 3 shows items that are presently recycled, together with some comments on the items themselves or on the recycling process.

Table 3. Recycling of materials found in MSW (Tchobanoglous *et al.*, 1993).

Item	Remarks
Aluminium cans	Industrial infrastructure ready to accept recycled Al
Paper and cardboard	Large quantities in landfills. Many types.
Plastics	Seven codified types by the Society of the Plastics Industry
Glass	Good reuse and recycling opportunities
Iron and steel	Recycling depending on market opportunities
Non-ferrous metals	(Al, Cu, Pb, Ni, Sn, Zn, etc.)
Yard wastes	Similar to the OFMSW
Organic fraction of MSW (OFMSW)	See more details in Section "The recycling of the OFMSW"
Construction and demolition wastes	Recycling depending on market opportunities
Wood	Minor component
Used tyres	Opportunity to waste-to-energy conversion
Household batteries	Separate collection extending. Recycling difficult
Lead-acid batteries	Good recycling market

As mentioned in the introduction, overall energy savings can arise from recycling. The issue of energy cost savings associated with MSW recycling has been addressed from an economic point of view by Lea and Tittlebaum (1993). Energy savings result if the energy required for collection, separation and treatment of reclaimed wastes along with subsequential processing is less than the energy used in originating and processing primary material and disposing of wastes. After analysing different scenarios, their results indicated that plastic should be not recycled for maximum energy recovery for two reasons: (a) plastic has to be very homogeneous to be recycled, and

(b) it is near the break-even energy costs for recycling when compared with feedstock prices. Based on this and without determining whether waste incineration is a desirable option, they concluded that waste-to-energy conversion is necessary if energy recovery and landfill volume minimization are to be met simultaneously.

Table 4 shows several issues for improving the recycling of MSW. Taking into account the above-mentioned study by Gilnreiner (1994), however, there may be some problems associated with the imposition through public relations and laws of recycling policies, and thus they are likely to fail. This is also in agreement with the examination of market incentives to encourage household waste recycling programme carried out by Reschovsky and Stone (1994), who state that efforts to encourage widespread recycling by imposing high quantity-based fees or by stringently enforcing mandatory recycling without, at the same time, providing a convenient means for households to recycle may be both unpopular and ineffective.

Table 4. Issues for MSW recycling (In addition to the educational issues shown in Table 1).

Adoption of tax-collection incentives
Compulsory use of recycled products in public institutions
Compulsory use of organic-waste-derived composts in public soils
Funding research and demonstration sites aiming towards recycling
Taxing energy and resources

Today the greatest inconvenience for the success of recycling programme is probably the lack of the infrastructure required for processing and marketing the recycled materials. In addition, there are several open questions that should be carefully studied before taking definite actions. As a first example: is it advisable to add a deposit fee for beverage containers? On the one hand it reduces the number left on curbsides but, it could discourage recyclers by diminishing the profits of the whole recycling process, because the lack of one of the more profitable components, namely glass. Another example concerning the type of plastics in some situations: Should they be made more biodegradable or should they be made more recyclable? From the energy point of view it seems that they should be more biodegradable (Lea and Tittlebaum, 1993), but other aspects such as the direct recycling of goods should also be considered. And finally, a third similar one: How does minimal not-recyclable-packaging compare with a heavier but recyclable-packaging? All these questions should be borne in mind in a holistic approach, considering an integrated waste managment system before setting standards, making appropriate Life Cycle Analysis (LCA), strengthening markets and even, educating consumers. In addition to this, optimal recycling efforts, and in general, optimal MSW management policies are likely to vary considerably from community to community.

A SUSTAINABLE APPROACH FOR MSW REDUCTION AND RECYCLING

Within the European Union (EU), only about 8% of MSW is recycled at present. A small portion (5%) of the organic fraction is processed as compost. Such low values reflect the fact that these materials are at the end of the cycle, their nature is very heterogeneous and, consequently, collection costs are high and the product final quality is usually low. On the other hand, selective collection is bound to have many difficulties, like those presented in Table 5, concerning the logistics of an extensive recycling of components present in household wastes. The question would be: How many materials should be classified at home and how many in the plant? In developed countries, it is fair to say that there is already a clear picture of the limits of what can be done, in an effective way, with residues. Even if all the measures suggested in Tables 1, 2 and 4 are applied successfully, a maximum around 50% could be achieved (Gilnreiner, 1994). An exponential trend towards this limit can be expected, but this is obviously a non-sustainable approach. Even in the case of a total absence of waste—which would minimize the use that is made of existing resources—they would still be depleted. How can the resources be recovered? First of all, and in order to surpass today's recycling limit, profound changes are needed in the manner of operation of productive systems.

A valid policy for a global strategy of waste reduction is ecological product design. The term "ecological design" for a product implies that it should not require disposal treatment processes. In fact, present policies in the European Union (EU) tend towards implementing the practice of clean technologies, which is a first step towards the adoption of the "ecological design". Existing regulations regarding ecological labelling are a clear example of such policies. In those rules, and making reference to products and processes, it is stated: "...they must guarantee a high level of environmental protection, and must be based, whenever possible, on the

Table 5. Questions on the limits of an extensive recycling of components of MSW.

Maximum number of bags allowed per household

Maximum number of different containers found on streets to deposit the sorted wastes

Maximum number of compartments per truck to collect the sorted wastes

Maximum number of trucks and their frequency

New materials, such as composites

---

use of a non-polluting technology, and when pertinent, must reflect opportunities for a maximum extension of the product life." Following the requirements for the EU-ecological-label, efforts in product design should be directed towards less toxic products (thus minimizing the environmental impact), more amenable to recycling (thus facilitating reprocessing), more easily eliminated (for example, more biodegradable) and longer lasting (decreasing of the global volume of residues). However, the "ecodesign", as it is also called, is something else. It constitutes a real technological challenge and at the same time, it constitutes a political one, because it is one of the most advanced strategies for preserving the environment.

In industrial processing, ecological product design should be the starting point for a long-term environmental policy. Up to now, present industrial policies have been aiming towards the minimization of requirements together with maximization of product profits. Moreover, distribution logistics have been also optimized. Environmental care has been included in an attempt to achieve the minimal allowable impact of the phases involved in the industrial activity, from production-distribution to consumption. In synthesis, the industrial concern for environment has been, and, to a large extent, still is, the part of the cycle that goes from production to consumption. The generation of wastes is considered as the price that a technological society must pay for the high levels of comfort achieved. Obviously, this trend cannot continue indefinitely, as it is not a sustainable policy to convert resources into residues. An attempt to reverse this trend should be made as soon as possible, that is, conversion of residues into products. However, this ambitious goal requires a true revolution in consumption habits.

Such a change should start at the very moment in which considerations are made on the conventional steps of product design such as productive process, product distribution and product life cycle. Additional considerations towards the logistics of used product recuperation, its eventual disassembly, its specific recycle and the proper utilization of the recycled product, are required at this early stage of design.

Such a concept would involve a substantial change in waste management, which could not employ the old scheme based on the recycling of used materials (aluminium, paper, glass, etc.). Recycling should be more ambitious and incorporate, in an integrated way and within a homogeneous logistic system, the concept of goods (and even services) recycling. Table 6 shows some examples of the recycling conventional procedures compared with the new ones.

The first steps have already been taken in Europe during the last few years. Significant actions can be observed in the research policies of large industrial firms (for instance, the German car industry already has a network for car recycling in operation), in aspects of regulations, as pointed

Table 6. Some examples of the new concept of recycling.

Conventional Scheme	New Scheme
Recycling of cardboard	Recycling of packaging
Recycling of lead	Recycling of batteries
Recycling of glass	Recycling of bottles
Recycling of iron	Recycling of refrigerators
Recycling of plastic	Recycling of packaging

out above, and in public research programmes. For example, these concepts appear in item 4 of the chapter on technologies for protecting and rehabilitating the environment of the R&D (1991-1994) European Union Program. This item points out, among others, research topics such as: the recycling of printed circuits, of electronic equipment, of household appliances, etc. In other words, emphasis is given towards research priorities which cover the whole life cycle of the product and the recycle concept is given in terms of applications and not materials.

The introduction of a concept such as the selling of "utilization of goods" instead of "selling the goods" is also a key issue in this new recycling philosophy. This concept would imply the technical development on utilization optimization rather than production optimization. Thus, the following strategies are aimed towards this goal (Stahel, 1992):
a) Prevention engineering: systems, products and component design with an unlimited life and with minimum or no maintenance required.
b) Design of adaptable systems: versatility of a system to adapt easily to new technologies, instead of being totally replaced by a new model.
c) Reliable design: operability and control incorporated within alarm and control actions.
d) Built-in protection products: which practically prevent any damage by direct action of the user (i.e., limits to motor speed).
e) Technical standardization of components.
f) User standardization of man-machine interfaces.
g) Training of operation and maintenance engineers at the highest degree.

A clear example of this tendency is the "Töpfer" law in Germany related to packaging materials and which has been in operation for more than three years. Such a law will make obligatory, fully by 1995, for producers, distributors, and merchants to reuse and recycle such materials from 64-72% depending on its composition. Moreover, new regulatory proposals are being studied for other products, as for example, refrigerator carcasses and newspapers.

Within the EU, there have recently been a series of regulations which follow the policy of "the one who pollutes, cleans", and even more "the one who assembles, dissembles". Firms that manufacture goods are responsible for them and should take the duty of disposing of the residues generated, not only during production, but also during consumption. These statements clearly show that the policy is to ascertain responsibilities beyond production to primary producers and not to society as a whole. Regulations are still under discussion, but it seems that the trend is to reduce the use of sanitary landfilling from about 60% to no more than 10-20%. Incineration will remain close to the present figure of 30%, while the rest will have to be either recycled or stabilized through composting.

To conclude, in Japan, the new recycling law has also the same goals and provides a legislative framework for guiding Japan toward becoming a recycling-oriented society. The law prescribes that the cooperation of industries, consumers and local governments are fundamental to the goals set by the specific enforcement ordinances (Yumoto, 1992).

THE RECYCLING OF THE ORGANIC FRACTION OF MUNICIPAL SOLID WASTE

The recycling of the organic fraction of MSW (OFMSW) involves the return of the organic matter, together with nutrients and micronutrients to the soil. To this end, this organic matter has to be previously stabilized by the so-called composting process. Aerobic composting is a process carried out by microorganisms in which the organic matter is converted to a stable humus-like material known as compost. Because it is an aerobic process, air has to be supplied to the system (there are several ways to do this). The time needed to achieve a completely stable compost depends on the characteristics of the incoming organic matter and the type of composting method used. Typical ranges are from ten days to three months for the fermentation phase and approximately one additional month for maturation. The first phase of aerobic fermentation can be advantageously substituted by anaerobic digestion.

INTEGRATION OF TECHNOLOGIES IN RECYCLING OFMSW: ANAEROBIC DIGESTION AS A KEY STEP FOR MATERIAL AND ENERGY RECOVERY

Contrary to what is believed by many engineers and practitioners, anaerobic digestion is a well-developed and mature technology which can guarantee its application without technical problems to the treatment of the OFMSW (Cecchi *et al.*, 1993; Edelman and Egeli, 1993). Anaerobic digestion produces a stable effluent, faster than the traditional aerobic fermentation, without the generation of obnoxious odours. This step is very interesting because of the recovery of energy: Anaerobic digestion by the action of microorganisms in the absence of air releases a gas, with around 65% methane content and 35% of carbon dioxide, which can be used to produce either heat and electricity. This energy can be utilized by the whole OFMSW treatment plant.

However, it should be emphasized that anaerobic digestion cannot be employed to produce a completely stabilized, mature compost. The effluent from the anaerobic digester, still cannot be applied to agricultural soils. Problems not only arise mainly from the phytotoxicity remaining, but also from the possible presence of pathogenic microorganisms (if low range temperature, around 35ºC, is used in the process). In addition, it is a very moist product that cannot be transported economically and, in some circumstances, if applied to some soils, can easily pollute underground water. Nevertheless, if a composting maturation phase is applied to this fraction, all these problems disappear. An aerobic finishing step removes excess moisture and phytotoxicity cheaply (Pavan *et al.*, 1992).

If the integrated process is compared with the conventional-single composting process, in terms of energy required, figures change, from about 30 kwh/ton for a single aerobic process (O'Keefe *et al.*, 1993), to a situation of excess energy due to biogas production during the integrated process.

As an example, in a recent paper the response of the thermophilic and mesophilic anaerobic effluents to a composting step was studied (Vallini *et al.*, 1993). Evaluation of the strategy adopted was carried out in terms of the performance of aerobic stabilization and economic considerations related to energy and the water balance of the process. The paper concluded that previous anaerobic treatment is a very attractive option whether performed at mesophilic or thermophilic temperature conditions (although the high range of temperature gives slightly better results). Its economic feasibility is guaranteed, especially when the integrated process operates with OFMSW selected at source. The additional fixed investment required is not significant, in comparison to the energy recovered, which can be utilized in the rest of the plant. Figure 1 shows the evolution of the most important composting parameters during the aerobic biostabilization carried out with the thermophilic anaerobic effluent. Anaerobic digestion of OFMSW was carried out in eight days of hydraulic retention time, whereas aerobic stabilization was accomplished in around three weeks, as can be seen in this figure.

A SUSTAINABLE APPROACH FOR THE RECYCLING OF THE ORGANIC FRACTION OF MUNICIPAL SOLID WASTE

Success in composting depends on product acceptability in the existing market. A generalized acceptance of compost relies on its quality. This is a

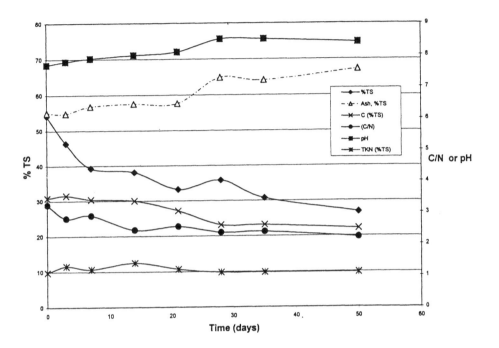

Fig. 1. Evolution of the indicated composting parameters during the aerobic
biostabilization of a thermophilic anaerobic effluent.

key point and quality control assurance must be implemented in a successful composting operation. Mechanical sorting always leaves inert material in the composted product, usually very small heavy metal particles from domestic products (paints, batteries, etc.), small particles of glasses, plastics, etc. Due to such problems, compost produced in plants with mechanical sorting facilities has a low quality and is rejected for extensive agricultural applications. It usually ends up as a cover in sanitary landfills for MSW. Obviously, this practice reduces the effective volume of the landfill but wastes an already processed and stabilized material. This is not, of course, a sustainable approach, even from a strictly economic point of view, because, as it is already happening in the USA, landfill costs are increasing drastically. Hence, the objective of recycling the OFMSW to produce a compost able to compete in the open market for agricultural uses, can be accomplished only if a high quality compost is produced. On the other hand, these characteristics of high quality can only be met by using high quality raw materials (The phrase "garbage in, garbage out", also applies for composting processes). The easiest way of obtaining "a high quality raw material" to be composted, is by using the OFMSW sorted at source. Selective collection is increasing every day and, this option will thus tend to be progressively more feasible. As a significant portion of the OFMSW comes from restaurants, fruit and vegetable markets, large establishments, etc., preventive measures should be taken in order to facilitate sorting of the non-biodegradable parts.

Finally, it is interesting to point out that the OFMSW coming from source selection enhances the interest of the integration of anaerobic and aerobic stabilization procedures compared with the straight stabilization through aerobic composting (Cecchi *et al.*, 1992). In fact, during composting, oxygen transfer is the limiting factor. Oxygen has a low water solubility and the moist solid particles are covered only with a very thin water later (1-2 mm). Therefore, anaerobic conditions are easily established, causing problems in the aerobic process and producing offensive odours. In consequence, anaerobic digestion is a recommended pre-treatment for these wet solids (the moisture content is above 80%) coming from a source separation.

SUMMARY AND CONCLUSIONS

Taking into account the environmental impact of technologies for municipal solid waste (MSW) management, a hierarchy can be established, in which source reduction would occupy the first place and, recycling, the second. The success of both approaches can be improved with political, technical and educational measures. However, there are presently some open questions that should be well studied before taking definite actions in any direction.

From a long term point of view, both approaches, reduction and recycling, are limited by ethological and social aspects. To overcome them, profound changes in product design technology are necessary. Goods should be produced for a very long life, enhancing the engineering of "utilization" over the engineering of "production". A major technological challenge for the next century is how residues can be converted into resources. Some steps in this direction have been taken in some legislations and in some large companies. Major changes are expected for the next decade in the way in which companies deal with their products at the end of their "first" utilization.

One of the components of MSW is the putrescible material. One sustainable way to recycle it is through composting. The composting process can be enhanced with a previous methane-producing step, which reduces the processing time and yields energy. In order to be able to apply the compost to the soil and thus, to recycle it, it should be of the highest quality.

This implies that a source selection of the organic fraction should be carried out.

REFERENCES

Anderson, D.D. and Burnham, L. (1992). Towards sustainable waste management. *Issues in Science and Technology* **9**: 65-72.

Cecchi, F., Mata-Alvarez, J. and Verstraete, W. (1992). Memorandum of the international symposium on anaerobic digestion of solid waste hold in Venice, April 1992. *7th International Recycling Congress*. October 28-30, Berlin, pp. 595-600.

Cecchi, F., Mata-Alvarez, J. and Pohland, F. (1993). Editors. Anaerobic digestion of solid waste. *Wat. Science and Technology* **27(2)**: 1-273.

Edelmann, W. and Egeli, H. (1993). Combined digestion and composting of organic industrial and municipal wastes in Switzerland. *Wat. Science and Technology* **27(2)**: 225-238.

Gilnreiner, G. (1994). Waste minimization and recycling strategies and their chances of success.. *Waste Management and Research* **12**: 271-283.

Lea, R. and Tittlebaum. (1993). Energy cost savings associated with municipal solid waste recycling. *J. of Environmental Engineering* **119(6)**: 1196-1216.

O'Keefe, D.M., Chynoweth, D.P., Barkdoll, A.W., Nordestedt, R.A., Owens, J.M. and Sifontes, J. (1993). Sequential Batch Anaerobic Composting. *Wat. Science and Technology* **27(2)**: 117-125.

Pavan, P., Musacco, A., Mata-Alvarez, J., Bassetti, A., Vallini, G. and Cecchi, F. (1992). Combination of anaerobic and aerobic bio-treatments of the organic fraction of municipal solid wastes. Economic evaluation. *7th International Recycling Congress*, op. cit. pp. 681-689.

Reschovsky, D.J. and Stone, S.E. (1994). Market incentives to encourage household waste recycling: Paying for what you throw away. *J. of Policy Analysis and Management* **13(1)**: 120-139.

Sikora, M.B. (1990). A little retreading goes a lot of miles. *Resource Recycling* **9(12)**: 50-58.

Stahel, W.R. (1992). Re-use and re-cycling: Waste prevention and resource savings in utilization. Waste Management International, K.J. Thomé-Kozmiensky (ed.), EF-Verlag Für Energie und Umwelttechnik GmbH, Berlin, Vol. 1, pp. 196-204.

Tchobanoglous, G., Theisen, H. and Vigil, S. (1993). *Integrated Solid Waste Management. Engineering Principles and Management Issues.* McGraw-Hill, Inc., New York, USA.

Vallini, G., Cecchi, F., Pavan, P., Pera, A., Mata-Alvarez, J., and Bassetti, A. (1993). Recovery and disposal of the organic fraction of MSW by means of combined anaerobic and aerobic biotreatments. *Wat. Science and Technology* **27(2)**: 121-132.

Yumoto, N. (1992). The new recycling law of Japan and its implementation. *Waste Management International* op. cit., Vol. 1, pp. 9-16.

20

Forestry, Municipal and Agricultural Wastes for Fuels and Chemicals Production in Yugoslavia—Resources and Engineering Data

M.S. Todorovic, T. M. Stevanovic-Janezic*, F. Kosi,
M.G. Kuburovic**, G. Koldzic and A. Jovovic

*Faculty of Forestry and
**Faculty of Mechanical Engineering
Faculty of Agriculture, University of Belgrade,
Agricultural Engineering Laboratory for Thermodynamics
Nemanjina 6, P.O. Box 127, 11081 Beograd, Yugoslavia.

INTRODUCTION

At the end of the second millennium, reaching fascinating technical and technological breakthroughs in many domains, humanity is facing equally distinct and of the same magnitude, problems of resources exhaustion and wastes disposal. Concerns on the exploitation of natural resources under intensified technical and economical development at the expense of environmental quality are on the increase and the goal of growing importance is to reach sustainability—to meet the needs of the present without endangering the ability of future generations to meet their own needs.

Waste is a byproduct of practically any activity or production process. Consuming resources and producing wastes is the way the industrial as well as natural ecosystems operate. As industrial systems grow, resources as well as sites available for wastes disposal become more limited. The extracted materials in extreme cases amount to even more than 10 tons per person annually, as for example in the U.S.A. where approximately 94% of the extracted material is converted to waste, while only 6% is finalised into durable products (U.S. Environmental Staff, 1995; Todorovic and Kosi, 1996). In addition to many examples of clearly dissipative uses of resources, many products are degraded, dispersed to the environment, and lost from the standpoint of any kind of reuse or recyclability to food, fuel, fertilizers, etc. Thus, the generation of solid wastes including toxic follows the same pattern as energy and air pollution-worsening as society grows richer. The enormous increase of the most developed nation's productivity and economic growth is based on technologies and consequently resources exhaustion. Hence, today are the most developed countries with the most increased industrial systems, regarding the resources exhaustion and wastes production, in a most critical stage. However, the reality is different.

Only a few distinct countries and nations are crowned by wealthy achievements and economic growth, but resources exhaustion and wastes generation and disposal are consequences spread nearly uniformly around the world, and it is now the Earth and Humankind as a whole at a critical point in environmental and economic policy making, to promote and implement environmental technologies processes that will minimize pollutants and recycle wastes, whenever it is possible, not only at the site of waste production, but even within the process, internally.

WASTE BIOMASS AND RESIDUES

Significant R&D progress has been made in the world in the area of environmental technologies and wastes production minimization, management and in situ utilization. There are also institutions in Yugoslavia engaged in similar R&D activities. A movement from research and development towards the market by the engineering applications and modern utilization of wastes for energy, chemicals and different types of recycled material production, requires reliable information. Engineering firms interested and ready to participate in technology transfer outline needs for more effective and efficient collection, comparison, critical evaluation and exchange of available R&D results and engineering data. There are also potential investors in Yugoslavia who need reliable information about efficient and cost-effective systems, design criteria and predictions of system's performance and relevant economy.

There is significant production of biomass and various types of biomass wastes and residues in Yugoslavia (Table 1). A bioenergy programme is currently in preparation, that will promote the use of waste and biomass as biofuels. The energy production by the utilization of waste and residual biomass is one of the most implemented means of exploiting renewable energy in Yugoslavia (Todorovic *et at.*, 1997). Different technologies are commercially available, including different types of fluidized bed combustion processes, biogas and biodiesel fuel production. Relevant data on quantities and energy potentials are given in the Table 2. However, the

Table 1. Biomass wastes and residues quantities and thermal energy potential in Yugoslavia.

Biomass Source	Amount 10³ (t/year)	Thermal energy equivalent 10³ (TJ/year)
Firewood	2195	27.47
Wood and forestry residues	84	1.05
Agriculture residues	7660	96.5
Liquid livestock waste	4000	19.9
Total		144.92

Table 2. Types and sources of forest biomass residues.

Source	Type - description of residues	Yield (%)
Forest operations	branches, twigs, foliage, stumps, roots, low grade logs (e.g. decayed)	38
Sawmilling	bark, sawdust, trimmings, split wood, planer shavings, sanderdust	37
Plywood production	bark, core, sawdust, lillypads, veneer clippings and waste, panel trim, sanderdust	43
Particleboard production	bark, screening fines, panel trim, sanderdust	45
Furniture production	cutting residue, sawdust, sanderdust	69.5

accuracy of potential estimation is limited, because presently the most of the biomass and wastes use is not occurring in large-scale combustion, facilities. Additional problems in analysing the status of biomass processing, biofuels market conditions and the conversion technologies presents the fact that the difference between waste and biomass is often difficult to determine.

WOOD RESIDUES FOR ENERGY AND CHEMICALS PRODUCTION IN SERBIA

Residues Quantities

Serbia has 2,422,200 hectares of forests, half of which are state-owned. This presents 27.4% of total Serbia area and the current plan is to reach 41.6% of forest covered area. Total wood volume contained in Serbia forests is estimated to be about 239 million of cubic metres, 57% of which is state-owned and 43% is in private forests. State-owned forests are managed by public enterprise SRBIJASUME (SERBIA FORESTS). Annual increase of wood volume is estimated to be 6.2 million cubic metres, of which 56% is produced in state-owned forests and 44% in private forests. Average annual increase of wood volume in cubic metres per hectare of forest is 2.6 (3.3 in high forests and 2.4 in low forests).

It is estimated that 63% of the wood biomass felled in the state-owned forests, and 88% in private forests in Serbia is inconvenient for further mechanical processing. Wood and forest residue, which might conditionally be named wood and forest wastes, could potentially be utilised for cellulose fibres, fibreboard and particleboard production, as well as for construction material and firewood, while technical foliage, comprising foliage, smaller branches and twigs, could be used for essential oils production, as a source of chlorophyll and vitamins (vitamin C and β-carotene) and other chemicals, as well as protein rich animal feed.

The residues generated by forest products' industries fall generally into two categories: those resulting directly from harvesting and extracting the logs from the forest and those generated from the process of manufacturing of wood (forest) products themselves. General description of the residues generated from forest operations and wood processing is given in Table 2 (Stevanovic Janezic *et al.*, 1994).

Utilisation of Wood and Bark Residues for Energy Production

Wood residues available for energy purposes are characterised by differences in regard of the chemical composition, morphological structure, moisture content and size. Limiting factors in the utilisation of wood residues for fuelling are problems related to collecting, storing and preparation of wastes generated from various stages of wood processing. Different percentages of various types of wastes are utilised for fuelling.

Transition of ownership, which has been taking place in Serbia, has influenced the conversion of larger-scale mills into many much smaller ones. Smaller mills are more flexible to the demands of internal and external markets. Some of them are situated in urban areas, which have to cope with the problems of both energy supply and waste disposal. Therefore, it is of particular importance to develop proper utilisation of wood and bark wastes in fuelling units of smaller capacity encompassing wood processing mills and handicraft workshops situated in urban areas.

As can be seen from Table 1, high quantities of bark are remaining as residues from various forest products industries. It is estimated that sawmilling and plywood production in Serbia yield approximately 100,000 m³ or 43,000 t of bark, while pulp production yields 30,000 m³ or 13,000 t of bark annually (Stevanovic Janezic *et al.*, 1994).

Determined higher heating values of non-extracted barks from beech, oak, spruce and pine range between 17,000 and 22,000 kJ/kg, which is comparable to higher heating values of respective woods of the same species. Higher heating values of woods and barks of selected native hardwood and softwood species of Serbia are presented in Table 3.

Table 3. Higher heating values and lignin content of barks and woods.

	Higher heating value (kJ/kg)			Lignin content (%)	
Species	bark*	bark**	wood	bark	wood
Fagus moesiaca		19500-20000		25.12	24.58
Quercus patraea	17296	17000	19580	16.19	23.54
Picea abies	21220	20385	-	22.76	28.00
Pinus sylvestris	22171	19220	21866	25.10	26.96

*non extracted; **extracted.

Light Briquettes

Briquetting procedure developed at the Faculty of Forestry, University of Belgrade consists of the following five steps as follows: (1) preparation of pulp from recycled paper which performs the adhesive function, (2) weighing of wood and bark biomass; (3) homogenisation of the briquetting mass, (4) pressing and (5) sewage of the briquetting mass and air seasoning of the briquettes. This procedure yields products with excellent handling and storage characteristics, of higher heating value per volume than firewood, of low ash and charcoal content and high volatile content. Light briquettes production and utilisation can contribute to the reduction of transportation costs as well as to the improvement of combustion efficiency.

The briquettes produced by the described procedure are sufficiently compact and water-resistant. Cohesiveness of the bio-briquette is achieved by the addition of groundwood pulp or recycled paper fibres. The described mixture of materials is almost saturated with moisture, and it is shaped in moulds by low pressures (three to five bar) application. During compressing, free evacuation of excess water from the mould is provided. The physico-chemical bonding of different materials in briquettes is taking place gradually and is continued throughout the air drying process, even after the moulding is finished. During the summer, drying of bio-briquettes by natural convection of ambient air is sufficient. During the winter period, forced air drying must be provided. Because of its simplicity, the described procedure is also applicable in households and in small-scale enterprises.

The advantages of the light briquettes production procedure presented here can be summarised as follows: 1) utilization of heterogeneous waste materials is possible, such as sawdust, conifer needles, bark (including particles coarser than 10 mm), pulp and paper waste, all these materials with different moisture contents; 2) free outdoor storage of these briquettes is possible, since swelling or splitting of layers, or structure decline was not observed, even for the increased moisture contents.

The temperature in the combustion chamber has great influence on the ignition delay time, but its influence on combustion time of volatiles is not significant. Results of compressibility measurements are encouraging, indicating that the densification of briquettes could be feasible. Preliminary experiments indicate that the briquettes maintain form and consistency upon charcoaling, which could open a new area of light briquettes application (activated charcoal).

Utilisation of Forest Residues for Essential Oils' Production

Conifer forests residues such as twigs and foliage (needles) represent an important potential raw-material for the volatile needle oils production. The feasibility of the essential oils production from the foliage residues from conifer forests managed by SRBIJASUME was in focus of our recent research project (Stevanovic Janezic et al., 1997). We studied the steam distillation of foliage from native pines : Black and Scotch Pine (*Pinus nigra and P. sylvestris*), Norway and Serbian Spruce (*Picea abies* and *Picea omorika*), White Fir (*Abies alba*) and Common Juniper (*Juniperus communis*). The equipment required for steam distillation is simple and the mobile systems available make small-scale production commercially feasible.

There are 111,943 hectares of conifer forests in Serbia which are managed by public enterprise Serbia - Forests. This public enterprise for forest management (JP SRBIJASUME, Belgrade) is directing and managing state owned forests and forest lands which encompass 1,373,500 hectares, of which 1,117,700 hectares are forests. Total wood volume contained in these forests is 135.66 million of cubic metres. Total annual wood volume increase amounts to 3.98 million cubic metres or 3.56 cu m per hectare. Total area covered by conifer forests consists of 80,582 hectares afforestated with pines (Black and Scotch Pine mainly), 27,020 hectares with spruces (Norway and Serbia Spruce mainly) and the remaining 4341 hectares are afforestated with different firs, junipers, other pines and spruces from those mentioned and other less abundant conifers. Forest resources of Serbia contain important conifer endemic species characteristic for the Balkan, such as Serbian spruce *Picea omorika* and endemic pines: Macedonian Pine *Pinus peuce* and Palebark Heldreich Pine *Pinus heldreichii* which could be potential sources of valuable chemicals.

Essential oil obtained by steam distillation of the needles of Serbian Spruce *(Picea omorika)* has been studied by gas-chromatography to determine terpene composition (Stevanovic Janezic et al., 1993). As a rule, a higher yield of essential oils was achieved from the whole forest residues (twigs, branches and needles) than from needles alone. It is therefore recommended to base the inforest production of essential oils on the whole forest residues, i.e. technical foliage consisting of foliage, twigs, tops and smaller branches.

Before steam distillation, the appropriate pre-treatment of raw material is necessary and therefore grinding devices have to be provided at forest sites selected for volatile oils production. In field tests performed so far, the laboratory grinder, operating on steel rotary discs, was used to reduce the technical foliage to particle sizes 1-2 mm, prior to steam distillation. We are redeveloping the convenient grinder for the field use, which requires source of electrical energy provided at each site of field application.

Anatomical Studies of Conifer Needles

Cross sections of needles of the examined pines, spruce, fir and juniper have been studied anatomically in order to determine the properties of their resin ducts. Shape, number per cross section and dimensions of the resin ducts were determined and correlated with the yields of volatile oils achieved in the laboratory study. The results of these studies are presented in Table 4.

Resin ducts are the most numerous in the needles of pines (6-8) but have the largest dimensions in fir (120 mm)—one order of magnitude higher than in the other studied conifers. Volatile oils obtained from the most important conifers of Serbia have been determined by gas-chromatographic analysis to consist of terpene compounds, among which the monoterpene hydrocarbons are dominant, as evidenced by data presented further in Table 5.

Chemical Studies of Essential Oils

Volatile oils obtained by steam distillation of technical foliage from the studied conifers have been examined by gas-chromatographic analysis to determine their terpene composition. Authentic monoterpene hydrocarbons, oxygenated monoterpenes and sesquiterpenes were used for identification of peaks in chromatograms. The maximum yields achieved in laboratory oil production from the conifer residues ranged from 2 to 9 ml/kg of fresh technical foliage. Exceptionally high yield was achieved for juniper fruits alone: 19.4 ml/kg, and the highest yield from technical foliage itself was achieved from fir technical foliage. These data have been correlated with the anatomical properties of its needles (Table 4). The obtained yields, versus the chemical composition of essential oils examined by gas chromatography, are presented in Table 5 (Stevanovic Janezic *et al.*, 1997)

Monoterpenes (hydrocarbon monoterpenes together with their oxygenated counterparts) contribute with 75-85% to the total identified terpene components, with the remaining percentage being covered by

Table 4. Anatomical properties of needles of *Abies alba, Picea abies, Pinus nigra, Pinus sylvestris* from the locality near Kraljevo, of *Picea omorika* needles from the locality at Jelova Gora, on Zlatibor and of *Juniperus communis* needles from the Tara mountain locality.

Property		Abies alba	Picea abies	Picea omorika	Pinus nigra	Pinus sylvestris	Juniperus communis
length (cm)	max	2.9	1.4	2.0	12.4	6.7	1.5
	✕	2.6	1.2	1.9	11.0	5.8	1.3
	min	2.5	1.0	1.3	10.7	5.3	1.1
width μm	max	1770	1440	2100	1352	1575	1470
	✕	1532	1335	1600	1283	1470	1298
	min	1350	1080	1400	1198	1120	780
thickness (μm)	max	720	975	990	806	900	510
	✕	630	913	945	748	850	464
	min	570	885	900	612	790	315
resin ducts							
		round			round to elliptic		
number	max	2.0	2.0	2.0	6.0	8.0	1.0
	✕	2.0	1.6	2.0	5.0	6.0	1.0
	min	2.0	1.0	2.0	5.7	5.0	1.0
width (μm)	max	165.0	99.0	153.81	65.0	71.5	120.0
	✕	123.6	37.1	141.51	42.9	52.5	88.5
	min	90.0	30.2	102.54	26.0	37.5	45.0
number of epithelial cells	max	21.0	13.0	17	9.0	11.0	13.0
	✕	18.6	9.0	14	6.9	8.4	9.8
	min	17.0	7.8	11	6.0	6.0	7.0
number of mechanical cells around resin ducts	max	23.0	18.0	20.0	11.0	14.0	14.0
	✕	17.4	16.0	15.2	8.2	9.4	11.6
	min	15.0	12.0	11.0	6.0	6.0	10.0

Table 5. Yields and compositions of essential oils obtained from technical foliage of *Pinus nigra, Pinus sylvestris, Picea abies* and *Abies alba,* from needles of Picea omorika and from technical foliage and fruits of *Juniperus communis.*

	Abies alba	Picea abies	Picea omorika	Pinus nigra	Pinus sylvestris	Juniperus communis needles and twigs	fruits
Y (ml/k)	8.9	1.9	1.6	4.7	4.0	4.15	19.4
trans-hexenol				1.0			
cis-3-hexenol						0.4	
monoterpenes							
santene	3.3	1.8	1.6	-	-		
tricyclene	1.5	0.6	0.8	0.1	0.9	2.0	1.6
α-pinene	12.4	9.6	5.0	54.2	36.5	36.9	23.3
camphene	12.3	9.2	7.3	1.0	4.3	0.4	0.2
β-pinene	27.9	9.6	0.6	2.9	8.3	18.9	25.6
myrcene	0.8	3.3	2.3	2.0	10.7	4.1	11.3
α-phellandrene	0.1	0.3	0.4	+	+	0.6	+
3-carene	-	0.7	0.1	+	+	0.6	+
α-terpinene	+	0.2	-	+	+	1.2	0.2
p-cimol	-	0.2	0.3	-	+	0.7	0.4
limonene/β-phel.	11.3	13.2	10.0	14.3	15.2	6.6	2.7
1,8-cineol	-	4.3	-	-	-		
γ-terpinene	-	0.3	-	+	0.1	2.1	0.3
terpinolene	0.6	0.7	0.9	0.8	0.4	1.6	1.1
total	70.2	54.0	29.3	75.3	76.4	75.7	66.7
oxydized monoterpenes							
borneol	4.9	4.9	-	0.5	1.6	0.5	0.2
menth-1-en-4-ol	-	0.7	-	0.1	+	6.1	1.5
α-terpineol	0.5	2.6	13.5	0.4	0.4	0.6	0.1
bornylacetate	8.4	10.9	29.4	0.8	0.3	0.4	0.7
longipinen	0.7	-		-	-		0.5
citronellylac.	-	1.1	2.3	0.3	-	0.5	
camphor			0.16				
piperitone			0.6				
geranylacetate			0.45				
total	14.5	20.2	46.41	2.1	2.3	8.1	3.0
sesquiterpenes							
longicyclene	0.3	-	-	-	-	0.7	
longifolene	0.4	1.1	-	-	-		
caryophyllene	3.8	1.9	-	3.1	1.1	1.0	1.8
humulene	1.6	1.5	-	0.6	0.4	0.6	1.9
germacrene-D	2.7	3.7	-	9.3	9.9	6.5	17.4
total	8.8	8.2	0.0	13.0	11.4	8.8	21.1

sesquiterpenes. Hydrocarbon monoterpenes dominate the composition of essential oils obtained from the studied conifers. The major component of the volatile oils obtained from the pines technical foliage was determined to be α-pinene. Its proportion ranged from 36.5 to 55.5% of all terpene components determined in essential oils from pine foliage.

β-pinene is determined as the major component of fir essential oil, while limonene/β-phellandrene is the major component of the spruce essential oil. Norway Spruce Essential Oil consisted of balanced quantities of α-pinene and β-pinene, camphene and bornylacetate (9-10%) and bornylacetate, an oxygenated monoterpene hydrocarbon, has been determined as the major component of Serbian spruce *(Picea omorika)*.

The application of essential oils from conifer residues is visualised mainly in relation to their fragrance qualities in production of soaps, bath foam perfumes and other cosmetic products. Terpenes are also notorious for their therapeutical properties and are applied in balsam production, nasal relief. Both α-pinene and β-pinene are characterised by the ease of conversion into other terpene structures by relatively simple chemical procedures. If the market value of the conversion products were sufficiently high, such conversions could be attractive commercially because α-pinene is available in high quantities from the examined essential oils.

It is interesting to note that essential oils obtained from technical foliage and fruits of Common juniper have comparable composition, with α- and β-pinenes as dominant components, the constituents are present in comparable proportions. The sesquiterpenes are an exception, as they appear to be more important in juniper cones than in foliage. As a result of conducted studies Srbijasume (Serbia forests) is planning to install a certain number of mobile distillery equipment at the locations at which sufficient amounts of forest residues, water and energy supplies are available.

Recycling of Residue from Essential Oils Production into Forest Ecosystem

Steam distillation of technical foliage represents a process which requires high quantities of water. After the green biomass (technical foliage) has been processed for essential oils' production, the residue which we can tentatively call final technological residue, or waste, is produced in a much higher yield than the targeted product, essential oil, which is commonly obtained in an amount of a few per cent of yield. Final technological residue after essential oils production, consisting of technical foliage and water solution left after steam distillation is completed, is produced in high quantities and therefore this technology requires further investigation and re-engineering.

The utilisation of the technological residue obtained after essential oils' production, consisting of technical foliage and water solution left after steam distillation, for composting contributes twofold to environmental protection: by composting the technological residue the chemical process of essential oils' production is closed, and by returning the compost to conifer forest soil the ecological balance of forest ecosystem is protected.

Conifer forests soils are more prone to leaching of minerals than soils of other types of forests, and therefore it is essential to return the biomass as entirely as possible to the conifer ecosystem. Therefore, it is our current task to study the decomposition potential of the technological residue left after steam distillation of technical foliage in comparison with foliage naturally felled at forest litter.

Decomposition of two foliage materials was followed in parallel by determination of mass loss together with other relevant parameters after six and twelve months from the initiation of the experiment, and the prognosis of intensity of degradation was given according to Olson's decomposition model (Olson, 1963; Waring and Schlesinger, 1985).

This investigation encompasses black pine *Pinus nigra*, as it represents over 50% of total conifer forest resource of Serbia. Comparison has been made between the chemical properties of technical foliage in a natural state and after steam distillation to estimate the starting potential for composting of the technological residue left after the steam distillation of foliage material.

The results of chemical analyses of two starting foliage materials used for composting experiments performed under laboratory conditions are presented here. Microscopical examination of the anatomical parameters of the pine needles before and after the steam distillation indicate the general decrease of the thickness of the needles—of chlorenchyma thickness in particular. Lignified tissues, such as epidermis, hypodermic, epithelial and vascular tissues appear to resist the process of steam distillation better. These tissues clearly demonstrate the presence of lignin by colour reaction with Wiesner reagent for lignin. The results of extractive contents determination in two foliage materials are presented in Table 6 (Stevanovic Janezic *et al.*, 1998).

As could be anticipated, the extractive content of distilled foliage material is decreasing, due to solubilisation in water and distillation (terpene

Table 6. Extractive contents of black pine foliage materials.

Sample description	% of Extractives in				
	benzen	ether	ethanol	hot water	1% NaOH
needles only	5.41	0.55	7.22	10.31	27.71
technical foliage	9.71	0.43	10.60	6.89	24.71
technological residue	9.51	0.34	5.25	5.63	27.55

components). Only relative increase of extractives content in 1% NaOH, which could be related to relative increase of aromatic content, was determined. Klason and acetyl bromide lignin contents were determined on two foliage materials after successive extraction and the obtained results are presented together with the results on %C and %N determinations in Table 7 (Stevanovic Janezic *et al.*, 1998).

Table 7. Lignin, carbon and nitrogen contents of two black pine foliage materials.

Sample description	%Klason	% A.S.	% A.Br.	% C	% N	C/N
needles only	14.82	0.20	17.04	44.10	1.00	44.10
technical foliage	14.20	0.55	19.11	51.66	0.91	51.66
technological residue	12.35	0.44	16.40	44.97	0.98	45.89

A.S.= acid soluble lignin; A.Br.= acetyl bromide lignin

Relative decrease of lignin contents is observed after the steam distillation which could indicate the presence of partial lignin degradation due to hydrolysis under steam distillation conditions. Lower lignin contents and lower C/N ratios in comparison with the technical foliage before steam distillation indicated that technological residue available after essential oil production from black pine technical foliage is a proper material for composting and returning to the conifer forest ecosystem. Water left after steam distillation should obligatorily be added to the composting biomass since it contains solubilized organic matter readily available for the growth of microorganisms. In this way, utilisation of conifer forest residues for essential oils production becomes a closed-cycle technology which is utilising its own residue for maintenance of conifer forest ecosystem equilibrium, while producing valuable new products for the market.

MUNICIPAL SOLID WASTES

Quantities and Composition

The municipal solid waste (MSW) is considered here as solid wastes originated in the municipalities, i.e., in households, administrative and educational institutions, manufacture and industries, tourist objects and trade, as well as the wastes from public areas (parks, construction and demolition wastes). In some cases in MSW, we include agricultural wastes generated by the various agricultural activities in suburban areas.

MSW are collected in Yugoslavia as a mixed stream, and landfilling is the most common type of MSW final disposal. There are in Serbia (one of the two federal state's units) more than 100 non-sanitary disposal sites which present potential sites of non-controlled fires and emission of gases including toxic components (CO, PCDD/F, PCB, and others).

The existing legislation and environmental policy recommend recycling of some materials (paper, glass and metal packaging materials) and a system of labelled containers for wastes separation and collection has been established, but is not yet effectively operational.

In addition to recycling, source-separation has attracted increased attention interest, because widely introduced source separation enables anaerobic digestion of MSW, results of which are encouraging. Co-digestion with other wastes, such as food-processing wastes and agricultural wastes like manure, is also an alternative option.

Data on MSW resources, i.e., quantities of wastes produced in the Yugoslav capital—Belgrade and its surroundings are as follows: in 1991, in Belgrade with approximately 2 million inhabitants, 287,508 tons of MSW had been collected (without the bulky wastes and rubbish), and in 1996 the quantity of MSW as approximately 311,000 tons (0.222, i.e., 0.223 tons/per inhabitant annually, respectively). During the period, the LTN economic sanctions had been in action (1992-1994), the economy destruction caused the total amount of wastes production to decrease to the level of 0.1820.201 tons/per inhabitant annually. In 1995, the amount of wastes started to rise again and as expected, economic recovery was followed by further increase in waste production. The estimation shows that waste production in Belgrade may reach the amount of 388,478 tons in the year 2000 (0.265 tons/per inhabitant i.e., 0.726 kg daily per inhabitant), and an amount of 521,134 tons in the year 2010. (0.323 tons yearly/per inhabitant or 0.885 kg daily per inhabitant) (Jovovic and Kuburovic, 1996).

The evidence shows that the waste composition has significantly changed in the last 50 years. The quantity of mineral components (ashes) has been decreasing and the content of paper, plastics and glass increasing.

With the rise in the standard of living, the morphological structure of the MSW is becoming similar to the MSW structure in the majority of the middle European countries, except the share of the organic components, which is bigger, to the detriment of synthetic materials, paper, glass and metals.

The composition of the MSW in Belgrade, according to Jovovic and Kuburovic, (1996) is as follows: the waste from disposal vessels (containers)- 79.28%, bulky waste—5.64%, construction waste—6.36 %, soil—8.47%, industrial waste in MSW—0.10%, and liquid waste-0.15%. More precise data are given in Table 8.

Low-temperature Oxidation

One of the important characteristics for the MSW disposal sites-landfills is

Table 8. Composition and density of MSW in Belgrade.

Component	Composition	Density
Paper (mixed)	16.56	22.27
Plastics (mixed)	7.61	2.67
Metals (mixed)	4.93	5.35
Rubber (mixed)	1.75	0.89
Leather (mixed)	1.59	0.89
Textiles (mixed)	1.71	1.78
Glass (mixed)	9.15	5.35
Food waste	36.73	32.97
Miscellaneous	19.97	16.93
Sum	100.00	89.10

Source: Jovovic and Kuburovic, 1996.

low-temperature oxidation (LTO) process (the oxidation process of the combustible wastes components at the ambient temperature), i.e. the MSW tendency to ignite spontaneously. LTO of wastes and explosions are processes which are also characteristic for the street sludge in the underground rain (not faecal) retaining canalization.

To assess the feasibility of a landfill ventilation and gas extraction a correct estimate of quantities and composition of disposed wastes and landfill gas is of major importance. The same data are necessary for other approaches to control the waste disposal sites as safe storage sites (waste bunker and silo) or to treat the waste by mechanical, thermochemical or biochemical conversion processes such as recycling, drying, composting, combustion, pyrolysis, or gasification.

Low-temperature oxidation of MSW has been studied experimentally. Relevant LTO experiments have been performed with samples that were

Table 9. Waste analysis results (after drying and grinding).

Component	wt. %
Proximate analyses	
Moisture	2.62
Ash	46.85
C-fix	6.77
Volatile	43.76
Ultimate analyses	
Carbon	58.30
Hydrogen	7.88
Nitrogen + Oxygen	32.44
Sulphur (comb.)	1.38
Lower Heating Value, kJ/kg	11304

collected on the waste disposal site near Belgrade during the winter period (Kuburovic, 1997).

The measurement data elaboration and determination of waste material feature relevant for the LTO process, and its inclination to autoignition is performed by the application of the Veselkovsky method (Kuburovic, 1997). Two methods of drying have been applied for samples pre-experimenting treatment: at the drying air temperature of 105°C, and by natural air convection.

Drying has been proceded with grinding to average particle diameter sizing of approximately 5 mm.

The range of moisture content values determined in samples was 27.91-43.80 wt %. The results of waste samples analysis are given in Tables 9 and 10 according to (Kuburovic, 1997).

Table 10. Waste ash analysis (after grinding, wt. %).

SiO_2	Fe_2O_3	Al_2O_3	CaO	MgO	SO_3	Fe_2O_5	TiO_2	Na_2O	K_2O
47.88	10.66	2.63	13.27	3.20	15.17	0.98	0.36	3.95	2.01

The experiments have been performed at three different temperatures within testing vessel: 20°C, 35°C and 50°C. A few of the oxidation rate constant (ORC) values (millilitres of oxygen consumption per gramme of sample and hour-ml/g·h) are presented in Fig. 1 (Kuburovic, 1997).

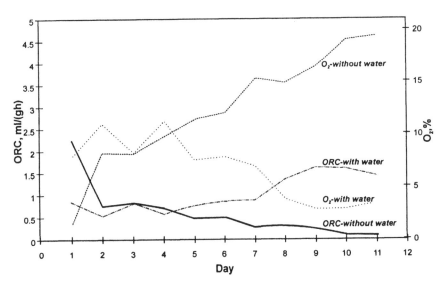

Fig. 1. Oxidation rate constant values during LTO at 50°C (sample II)

The investigation results show that 1000 tons of disposed MSW (a quantity collected daily in a city of million inhabitants) is able to consume 20,000 m³ of air in 24 hours. Thus, by order of the corresponding necessary amount of oxygen, landfill gas is recognized as a significant potential for utilisation as an energy source, if not, there is a serious need to protect - ventilate landfill (control of landfill gases), and placing special green belts (oxygen sources) around the waste disposal sites.

ANAEROBIC DIGESTION OF BIOMASS AND MSW

In addition to the biomass use by the combustion process, in Yugoslavia several biogas plants have been built and are in operation. Agricultural wastes, sludge from water purification plants and surplus products from industry, processed and biochemically converted by anaerobic digestion to organic acids, solvents and gases present tremendous potential for the production of solvents such as ethanol and butanol, volatile fatty acids used as acidulents, preservatives, flavouring agents or chemical feedstocks, and energy-rich biogases (McCaskey *et al.*, 1994; NIP Republic Min. of Sc. and Tech., 1995; Todorovic *et al.*, 1996). By coupling fermentative processes with microbial or enzymatic saccharification of biomass, wastes from lumbering, agriculture, food processing, and other sources can be sequentially converted into these saleable products. In addition to implementation results of anaerobic methane and alcohol fermentations are enhancement of waste material destruction and reduction of the environment pollution.

Both, pilot- and large-scale bioreactor systems were built in Yugoslavia, but their operational efficiency and productivity is far less than their potential. Functional imperfections and uncertainty in the continuous process control have often precluded maintaining sufficient long-term productivity rates. We attribute this to present limitations in the knowledge of microbiological, thermophysical and rheological properties of biomass and fermentation substrates, and to a lack of parametric data on process kinetics and the synergism between biochemical-, mass- and energy-transport processes.

Simultaneous fermentation and separation of an inhibitory fermentation product, being an innovative process is the focus of our study (NIP Republic Min. of Sc. and Tech., 1995). In a biparticle fluidized bed reactor containing immobilized cells increases productivity because potentially-inhibitory products can be removed easily from the system. Thus, integrating fermentation and product separation processes in a single reactor minimizes by product formation and reduce waste production at its source.

Mathematical modelling of bioreactors reveals that greatest emphasis has been placed on empirical relationships (Kosi and Todorovic, 1991; NIP

Republic Min. of Sc. and Tech., 1995; Todorovic *et al.*, 1996). The applicability of analytical models and related mathematical formulations has been hampered by the limited availability of experimental data on bio- and thermophysical properties of heterogeneous reactor contents mixed MSW and bioeffluents. In addition, the influence of different growth stages of immobilized cells on bioreactor phenomena, and likewise the influence of transport phenomena on microbial growth and activity, require more complete description under controlled, repeatable conditions. Done on both experimental and theoretical levels, such studies may possibly result in a new, more accurate form of correlation.

Thermodynamical and Rheological Data Determination

Thermodynamical and rheological properties of agricultural liquid effluents relevant for mechanical and biochemical processing including methods of their experimental determination have been studied (Kosi and Todorovic, 1991, 1995), and a sample of obtained measurements results is here presented. The intrinsic phenomenological aspects of the relevant property determination need improvements of their physico-chemical and measuring-technical characterization as well as corresponding analytical formulations. The intrinsic time and space dependent variability of the physical and chemical properties of the studied materials were confirmed. Their complex structure, distinct polyphasic heterogeneity and anisotropicity complicate material description and repeatable measuring characterization.

Specific heat capacity of samples of livestock waste slurries has been measured by the Bunsen "ice calorimeter". The specific heat capacity of the material is obtained through measuring of the enthalpy change by cooling of the sample until the temperature of thermal equilibrium of the system "sample-calorimeter" is achieved.

Experimental determination of heat conductivity has been performed using concentric-cylinder-measuring apparatus based on the one-dimensional steady state Fourier heat conduction law. Due to the heterogeneity of manure, specific problem in performing these measurements arises regarding the time elapse for the steady-state measurements and the tendency of solid particles suspended in the manure to settling and wall deposit. Relevant analysis has been made, the result of which shows that due to the specific rheological properties of manure and to the tendency of solid particles to form irregularly shaped clusters, in the narrow space between the cylindrical surfaces of the measuring device appears a "local" settling, i.e., clogging which prevents a vital jeopardizing of the assumed homogeneity of the sample.

Materials specific density has been determined by the direct measurement of volume V of a certain mass m of the sample previously brought to a desired temperature, assuming that intensive mechanical mixing during the measurement procedure guarantees the uniformity of materials sample temperature and concentration fields and hence the relevance of obtained density values as $\rho = M/V$.

Rheological model characteristics have been obtained by the determination of the correlation between the measured fluid shearing stresses and shear rates on the wall surface. Concentric-cylinder rotary viscometer of the Brookfield Synchro-Lectric type with four spindles has been used for the measurements. The rheological behaviour of the pseudoplastic fluids can be described by the Ostwald de Waele equation as follows:

$$\tau_w = K \cdot \left(\frac{\partial v_x}{\partial y} \right)^n \tag{1}$$

where K (Pas^n) is rheorogical consistency index and n rheorogical behaviour index.

The second order regression analysis is used to relate measured thermophysical properties to the temperature:

$$\pi(t) = a + b \cdot t + c \cdot t^2 \tag{2}$$

where π is values of thermophysical property and a, b and c are the constant coefficients. The obtained coefficient values are given in Table 11. The performed measurements of the rheological properties confirm that manure is a pseudoplastic fluid. A significant influence of the sludge volume fraction on the values of rheological parameters K and n is observed.

APPROACHING THE SIMULATION AND OPTIMIZATION OF WASTES PROCESSING CHAIN

Processing of waste materials is commonly described as multistage processing by a series of interrelated mechanical processes and physico-chemical reactions which are to be performed in different units of processing systems. Since the units of such a processing system are linked by mass and energy fluxes, the procedure of a system's mathematical modelling, with the aim of determining parameters relevant for designing the equipment and controlling distinct technologies and processes performance, is to be based on the sound physical description, accurate calculations and reliable parametric and physical and/or biophysical data. To approach a system's optimization, economic analysis of all subsystems and different scenarios of structuring technologies and technical units and subunits has to be included, and a powerful analytical method has to be established. It has been experienced that a general system of recurrent equations based on Belmopan's principle of dynamic programming may be a successful

Table 11. Experimental coefficient values for livestock waste slurry (for eq. 1).

		PIG			DAIRY	
C_R (kg/m³)		90	60	41	90	60
Heat capacity $c(t)=a+bt$	a (kJ/kgK)	3.9312	3.9534	4.0713	3.9214	3.9324
	b 10³ (kJ/kgK²)	0.7911	0.7713	0.5431	0.7869	0.7583
Thermal conductivity $k(t) =a+bt+ct^2$	a (W/mK)	0.6024	0.6013	0.5755	0.6246	0.5899
	b 10² (W/mK²)	0.4162	0.2988	0.2950	0.2078	0.2788
	c 10⁴ (W/mK³)	−0.3494	−0.1811	−0.1833	−0.0665	−0.1735
Density $d(t)=a+bt+ct^2$	a (kg/m³)	1031.1	1021.0	1014.1	1028.1	1018.7
	b 10 (kg/m³K)	−0.3721	−0.6404	−0.4193	−0.2193	−0.1095
	c 10² (kg/m³K²)	−0.4432	−0.3860	−0.4391	−0.4661	−0.4587
Rheological consistency index $K(t)=a+bt+ct^2$	a (Pasⁿ)	6.0165	0.7246	0.3593	12.7150	3.0110
	b 10² (Pasⁿ K⁻¹)	−4.5505	−0.5786	−0.2672	−6.6660	−2.0307
	c 10⁴ (Pasⁿ K⁻²)	5.0518	0.5833	0.2550	6.1296	1.2179
Rheological behaviour index $n(t)=a+bt+ct^2$	a (-)	0.4743	0.5667	0.5990	0.3470	0.4184
	b 10⁴ (K⁻¹)	−0.7093	−0.9415	−0.3984	−0.4165	−0.5848
	c 10⁶ (K⁻²)	0.7130	0.8923	−0.2596	0.1875	0.3645

tool to approach the optimization of the multistage wastes processing chain.

When considering multistage processes, it is always possible, taking into account certain presumptions, to define a system of equations for describing each stage of the process and to put them together into a mathematical model of the whole process. Applying such a principle of making a physical and mathematical model has many advantages especially concerning testing and numerical realisation of the model. Some parts of a mathematical model can be tested separately, and a model of one stage once confirmed, can easily be integrated into process systems for different purposes.

The method of dynamic programming Bellman (Bellman, 1971; Kosi and Tododrovic, 1991b) is a sequential decision making process, which represents a discrete case of optimum control problem (Bellman, 1971). A particular system for which an optimum is sought is broken down to a number of series of stages, and an optimal solution is obtained by making a sequence of interrelated decision procedures at different stages.

The most general form of a serial multistage decision problem for a non-cyclic structure system analysed is shown in Fig. 2 in which the stages are represented schematically by appropriately numbered rectangles, with arrows used to indicate inputs and outputs to various stages.

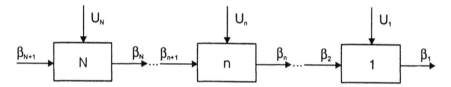

Fig. 2. General form of serial multistage problem for an non-cyclic system

Associated with n-th stage, variables, β_n and β_{n+1}, defined in the domain given by inequality constraints of the following type

$$\beta_n : \Phi_n(\beta_{n,l}) \leq 0, \; n=1,2,...,N; \; l=1,2,...,L_N \qquad (3)$$

are state variables of the n-th stage. These variables, which, in a general case, are multidimensional vectors with scalar physical values as "coordinates", present characteristics of the mass, energy and information flows between the stages. State variables undergo some changes in value at each stage defined by relationships

$$\beta_n = T_n(\beta_{n+1}, u_n), \; n=1,2,...,N \qquad (4)$$

which are state-transformation functions T_n. The decision (or control) variables u_n, $n=1,2,...,N$, represent the parameters (as temperature, pressure, concentration, etc.) that determine the contribution to the output β_n, $n=1,2,...,N$, of a particular stage. In the optimal control theory (Bellman, 1971), the problem of choosing (prediction) of the decision variables is solved by definition of the functions u_n in the domain U_n given by non-equations system of the following type

$$U_n = \Psi_n(u_{n,j}) \leq 0, \; n=1,2,...,N; \; j=1,2,...,J_N \qquad (5)$$

Relevant parameters of quality of a controlled system treated are evaluated by a scalar criterion functions g_n of the stage n of the form

$$g_n = g_n(\beta_{n+1}, u_n), \; n=1,2,...,N \qquad (6)$$

which are called control quality criteria or n-th stage return (objective function, as well). For the system as a whole, in a general form given the function

$$G = G(g_n(\beta_{n+1}, u_n)), \quad n=1,2...,N \tag{7}$$

is an overall objective function of the problem treated, whose extreme value is defined by the optimal state variables β_n and decision variables u_n. The equation (6) is possible to express in additive form (8), which is essential for the practical application of a dynamic programming procedure.

$$G = \sum_{n=1}^{n=N} g_i(\beta_{n+1}, u_n) \tag{8}$$

For the case, processed wastes are to be used as energy source, the overall objective function of the optimisation model is usually defined by "the present-worth" method . Thus the defined problem is typically solved recursively beginning with the last stage, computing optimum return $g_n{}^* = f_n$ for each possible entering state variable values at every stage. Of course, the decision variable u_n at any stage has an effect on the inputs to all subsequent stages in accordance with equation

$$f_n(\beta_{n+1}) = \max(g_n(\beta_{n+1}, u_n) + f_{n-1}(\beta_n)), \quad u_n \in U_n \tag{9}$$

known as Bellman's function (Kosi and Todorovic, 1991). So finally, the optimum return G^* for the system as a whole, is expressed as

$$f_n(\beta_{n+1}) = G^* \tag{10}$$

The first step in applying the described dynamic programming technique for optimization is suitable structuring of the whole chain as a serial, sequential process. The scheme of the multistage-municipal wastes processing chain, with codigestion shown in Fig. 3 can serve as an example in that sense.

Management and disposal of the growing quantities of waste materials is a task which is as difficult as it is essential in world's society today. Searching and finding acceptable scenarios of wastes management and disposal, reducing waste "leaks", preventing toxic wastestreams to enter environment, increasing wastes reuse, we can minimize disposal needs, and we can say that we are approaching quasi-sustainability (current phase).

To enter new frontiers of real sustainability, it is necessary to invent new, inherently environmentally, clean processing technologies. That is to be in the near future (next phase), not to be too late. However, in both phases, given the interwoven nature of environmental problems, systems approaches are essential if we are to attain sustainable development. Thus, in both phases, the transition to a system approach is an evolution-ary process of integrating the all relevant environmental areas identified above.

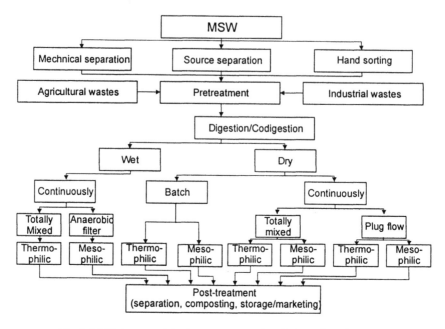

Fig. 3. Multistage-municipal wastes processing chain with codigestion.

REFERENCES

Bellman, R. (1971). Introduction to the Mathematical Theory of Control Processes, Vol. II, Non-linear Processes, Academic Press, New York/London.

Fundamental and Applied Researches Relevant for New and Renewable Energy Sources Development, NIP Republic Ministry for Science and Technology 1991-1995, Final Report, 1995.

Jovovic, A. and Kuburovic, M. 1996: The Possibility of Recovery of Material and Energy from MSW, Recycling Plant Project-Study, Report No. 503/707/97, Faculty of Mechanical Engineering, Belgrade.

Kosi F., and Todorovic, M. (1991). Experimental Investigation of Rheological Model Characteristics of Non-Newtonian Fluids, International Conference "Energy Efficiency 2000", pp. 253-264, Belgrade.

Kosi, F., and Todorovic, M. (1991). Mathematical Modelling and Optimisation of the Anaerobic Fermentation of Agricultural Wastes, KGH, No. 4, pp.52-57, Belgrade.

Kuburovic, M. 1997. Investigations of Low-temperature Oxidation Process of Solid Communal Waste, *Poljoprivredna tehnika*, Vol. 21, 1, 85-99.

McCaskey T.A., Zhou, S.D., Britt, N. Sarah, and Strichland, R. (1994). Bioconversion of Municipal Solid Waste to Lactic Acid by Lactobacillus Species, Applied Biochemistry and Biotechnology, Vol. 45/46, pp. 555-563.

Olson, J.S. (1963). Energy storage and the balance of producers and decomposers in ecological systems. Ecology44, 322-331.

Stevanovic Janezic, T., Bujanovic, B., Vilotic, D., Dajic, Z., Mitrovic, M. and Pavlovic, P. (1998). Significance of Lignin Content for Composting of Lignocellulosic Materials

Proceedings of the 5th European Workshop on Cellulosics and Pulp (EWLP), Portugal, p. 129-132.

Stevanovic Janezic, T., Bujanovic, B., Vilotic, D., Dajic, Z., Mitrovic, M. and Pavlovic, P. (1997). Recycling of Residue from Essential Oils Production into Forest Ecosystem, Proceedings of 5th Brazilian Symposium on the Chemistry of Lignins and other Wood Components,

Stevanovic Janezic, T., Danon, G., Bujanovic, and B. Dedic, A. (1994). Bark Utilisation for Energy Purposes, Proceedings of the 8th European Conference on Biomass for Energy, Environment, Agriculture and Industry, pp. 1607-1611, Vienna.

Stevanovic Janezic, T., Isajev, V. and Lange, W. (1993). "Needle oil terpenes of Serbian spruce from three different localities "Holz als Roh- und Werkstoff, 51: pp. 283-286.

Technology for a Sustainable Future (1995). A Framework for Action, U.S. Environmental Technology Strategy Staff, Alamos National Laboratory, New Mexico.

Todorovic M., and Kosi, F. (1996). Environmental Energy Technologies—New Energy Conversion Systems for Thermotechnics and Thermoenergetics, KGH, Vol. 1, pp. 45-54, Belgrade.

Todorovic, M., Gburcik, V., Djajic, N., Gburcik, P., Stevanovic-Janezic, T. and Kosi, F. (1997). National Information & Technology Transfer Network on New and Renewable Energy Sources, Proceedings of the European Congress on Renewable Energy Implementation, pp. 934-943, Athens.

Todorovic M., Boyce E., Kosi F., and Simic Lj. (1996). Anaerobic Bioreactors Producing Chemicals and Fuels—Innovations and Engineering Database, Proceedings of the 9th European Bioenergy Conference, Vol. 3, pp. 1560-1565, Copenhagen.

Waring, H.R., Schlesinger, H.W. (1985). Forest Ecosystems. Concepts and Management, 340 pp., Academic Press, INC.

Section 5

Country Case Studies of
Latin America

21

Solid Waste Management in Latin America and the Caribbean

Marcia Marques Gomes[1,2] and William Hogland[1]

[1] Department of Technology, University of Kalmar,
P.O. Box 905, SE-391 29 Kalmar, Sweden
[2] Department of Environmental Engineering,
Rio de Janeiro State University, Rio de Janeiro, Brazil

INTRODUCTION

According to the United Nations, in 1995 the Latin American and Caribbean (LAC) region had 482 million inhabitants. The population size of the LAC countries varies from less than 0.5 million to about 160 million inhabitants (Table 1). It is predicted that by the year 2010, this region will have around 604 million inhabitants, and that about 80% of the population will be living in the seven most populated countries.

During a 20-year period (1975-1995) the urban population in the LAC region increased by 80%. This increase led to a relatively sudden, high demand for public services. The complexity of the public services, which require qualified personnel and investments in a long-term perspective, makes sudden, high demands, which are difficult to meet. Small and medium-sized cities will probably face rapid growth in the near future requiring substantial investment and technical assistance. Population growth in the larger cities of the LAC region is showing a trend towards stabilization, but there will still be a high demand in terms of public services.

During the period 1990-1997, most of the national economies in the LAC region experienced a reduction in inflation rates, a significant income from external capital investments and a moderate expansion. The recovery in the level of economic activity that had begun in 1996 gathered speed in 1997, and the total domestic product of the region in 1997 was approximately US$ 1.3×10^9. However, in 1998 the economic activity has slackened, in large part as a reflection of the Asian crisis (ECLAC, 1998). Despite the economic growth in the region during the 90's, the percentage of the poor population, particularly in urban areas in all the LAC countries is still very high. Only in two of the Latin American countries analysed by ECLAC (Argentina and Uruguay) fewer than 15% of households live below the poverty line. Honduras has the highest level of poverty with about 50% of their households below the poverty line. In the Caribbean region the poverty rate varies between 12% and 42% of the population. This, together with a lack of proper educational programmes has, so far, placed a serious constraint on the implementation of financially self-sustainable waste management systems in the region.

Lack of information places a serious constraint on planning and project design. However, this problem has been addressed and information about solid waste management (SWM) in various countries in the LAC region is now available as a result of local efforts and international support. Several diagnoses and sector analyses have been conducted by the Panamerican Health Organization (OPS), the World Health Organization (WHO) and the Interamerican Development Bank (IDB) during the period 1995-1997 from which, most of the data presented here were extracted. The large diversity concerning economic, cultural and social aspects as well population size verified in LAC cities makes difficult to establish figures that represent the region. Therefore, the values presented in several tables in this paper should be taken as magnitude for parameters.

Table 1. Population in Latin American and Caribbean countries in 1995.

Population (millions)	Country
Less than 0.5	Antigua and Barbuda, Antilles Nerlandesas, The Bahamas, Barbados, Belize, Dominica, Grenada, Guadeloupe, Saint Kitts and Nevis, Santa Lucia, San Vincent and Las Grenadines, Surinam
0.5-1.0	Guyana
1.0-5.0	Costa Rica, Jamaica, Nicaragua, Panama, Paraguay, Trinidad and Tobago, Uruguay
5.0-10.0	Bolivia*, El Salvador*, Haiti**, Honduras*, Dominican Republic*
10.0-20.0	Chile, Cuba, Ecuador, Guatemala**
20.0-50.0	Argentina, Colombia, Peru, Venezuela
50.0-100.0	Mexico
more than 100.0	Brazil*

* Between 15% and 30% of inhabitants older than 15 years are illiterate.
** More than 30% of inhabitants older than 15 years are illiterate.
  Source: (Acurio *et al.*, 1997).

STRATEGIC ASPECTS

Most of the infrastructure in the LAC region has been dependent on external investment, which has been made mainly in the energy, transport and telecommunication sectors. Second to these, come water supply and sanitary services. Despite a recent increase in investments, SWM has not attracted external investment in the same degree as other sectors, and is still mainly dependent on national funding (Gomes, 1995).

As a consequence of recent neo-liberal political trends, many countries have implemented privatization programmes in various sectors, including SWM, mainly in large and medium-sized cities. The majority of the contracts are concerned with waste collection and transportation. More recently, third-party services and concessions have included final disposal facilities. In cases where the increasing participation of the private sector has functioned properly, a reduction in waste management costs was observed as a result of the competition between contractors. In the LAC countries, the reduction in costs was around 50% compared with 25%-45% reduction in developed countries following privatization (Acurio *et al.*, 1997). It should be pointed out that in many cities in the LAC region privatization programmes have caused social conflict, between municipal workers and the authorities since in many cases privatization has led to unemployment or salary reductions.

Various programmes and pilot projects have been developed in many LAC countries during the last few years, frequently with support (loans or grants) from international institutions. However, problems often arise during implementation. Most of the major plans developed for large metropolitan regions are seldom implemented, while pilot projects seldom last for more than few years. A problem identified in most studies is the lack of institutional, administrative and economic sustainability in SWM schemes in many LAC cities. Successful exceptions usually include the creation of small enterprises and private initiatives. To address the problem of sustainability, financial institutes, such as the IDB and the IBRD/World Bank, have established the development of a "master plan" which includes strategic planning to guarantee economic and institutional sustainability, as a precondition for loan concession.

TECHNICAL ASPECTS

Municipal Solid Waste Generation

The amount of household waste generated daily in the LAC cities varies from 0.3 to 0.8 kg/person (Acurio *et al.*, 1997). Other types of waste which together with household waste constitute municipal solid waste (MSW), such as commercial, institutional, industrial waste from small plants and public waste (from street cleaning), account for 0.07-0.4 kg/person giving a total amount of MSW of 0.4-1.2 kg/person/day (Table 2).

Table 2. Proportion of various components of MSW in the LAC region expressed as percentages (data from Acurio *et al.*, 1997).

Household waste	Commercial waste	Institutional waste	Small-industries waste	Public waste (from street cleaning)
50-75	10-20	5-15	5-30	10-20

The generation of MSW is associated with consumption patterns, and trends for poor and rich groups can be observed when the LAC countries are grouped according to the average income per capita (Table 3). When LAC cities are classified according to their population size, it can be seen that large cities may generate as much waste as the average for European countries (Table 4), the difference being in the composition of the waste.

Table 3. Daily MSW generation in LAC countries according to the average income per capita (data from CEPAL, 1995).

	Countries with lower income per capita	Countries with intermediate income per capita	Countries with higher income per capita
Daily waste generation	0.4 - 0.6 kg/person	0.5 - 0.9 kg/person	0.7-1.2 kg/person
Gross National Product US$/capita	< 500 - 999 (8 countries)	1000 - 2999 (10 countries)	3000 - 10000 (6 countries)

Table 4. MSW average daily generation of MSW per person in LAC cities grouped according to size compared to the average for the USA, Japan and European countries.

Average generation	USA	Japan	European countries (22 countries)	LAC: metropolitan areas and large cities > 2 million inhabitants (16 cities)	LAC: medium-sized cities < 2 million and >500,000 (16 cities)	LAC: small cities < 500,000 inhabitants (24 cities)
kg/person	1.97*	1.12*	1.03*	0.97**	0.74**	0.55**
Range			(0.69-1.36)	(0.54-1.35)	(0.51-1.20)	(0.30-1.20)

Source: * EUROSTAT (1994) and OECD (1993) data in White *et al.*, 1995. ** Acurio *et al.*, 1997.

Hazardous Waste

One of the most serious problems identified in studies conducted in the LAC region is the fact that hazardous and municipal wastes are handled together. The hazardous waste generated by medical institutions is estimated to be around 600 tons daily (0.5 kg/bed/day). However, from the quantitative point of view, hazardous waste from industries represents a more serious problem than hazardous waste from medical institutions. The metropolitan region of São Paulo State in Brazil, one of the most industrialised areas in the LAC region alone generates 554 tons of hazardous industrial waste per day. This represents almost the same amount of hazardous waste generated by medical institutions of the whole LAC region, 52% of which receives proper treatment and disposal.

Waste Composition and Characteristics

The composition of MSW reflects the income, the stage of industrial development and cultural aspects of a country or region (Table 5).

Table 5. MSW characteristics and composition (% by weight) in the LAC region, and the USA.

Region/ Country	Moisture content	Specific weight	Organic fraction (putrescible)	Recyclable fraction (% weight)				
	% weight	kg/m^3	% weight	paper	plastic	glass	metals	others
LAC[1,d]	35-60	160-250	40-70	6-25	3-20	1.6-8	1-7	1-5
USA[2]		100	7[a] +14[b] +7[c]=28	39	10	6	8	9

[a] Food waste [b] yard trimmings [c] wood [d] excluding Trinidad and Tobago
[1] Acurio *et al.*, 1997. [2] US EPA, 1995.

A higher organic fraction is found in LAC countries compared to that in developed countries. The waste moisture content is also higher than in developed countries. The same is observed for the bulk density: 200 kg/m^3 in the LAC region compared with 100 kg/m^3 in the U.S.A. Higher bulk densities are related to a lower proportion of recyclable material in the waste. An increase in the recyclable fraction, mainly plastic and paper has been observed in many cities in the LAC region and this has been interpreted as an indication of industrial development. Considering the average values of recyclable material in MSW, paper and cardboard (6-25%) is followed by plastic (3-20%), glass (1.6-8%), metals (1-7%) and textiles (1-5%). These values exclude Trinidad and Tobago whose waste composition is very similar to that in developed countries (a higher percentage of recyclable materials and a lower percentage of organic matter and lower moisture content).

Figure 1 shows the thermal characteristics of MSW and values required for self-sustained combustion.

The calorific value is directly proportional to the content of recyclable material and is an important parameter when selecting treatment options for waste. Although the calorific value may vary greatly, in the LAC countries the lowest calorific value found is around 4.2 MJ/kg, while in developed countries where incineration is more common, for example the U.S.A. the calorific value of waste is not lower than 11.7 MJ/kg.

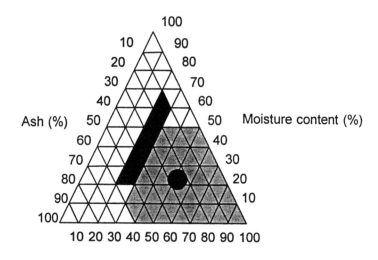

Area indicating self-sustained combustion
Typical values for many developing countries
Typical values for many industrialised countries

Fig. 1. Comparison of the thermal characteristics of MSW and those required for self-sustained combustion (Diaz *et al.*, 1996).

Collection

The number of collection trucks per inhabitant in LAC cities varies from 1/7,000 to 1/27,000. The Metropolitan region of São Paulo in Brazil presents the lowest ratio, namely one truck per 27,000 inhabitants (Acurio *et al.*, 1997). In the U.S.A., the average ratio is one collection truck per 4,000 inhabitants. Although the ratio of trucks to inhabitants is lower, large cities in the LAC region have a collection frequency of three times per week, while in the USA the frequency of collection is not more than one to two times per week. The collection of MSW in the LAC countries thus poses a considerable challenge to managers and operators. Under these circumstances, the collection of 95% of the MSW achieved in many LAC large cities is considered a surprisingly satisfactory result.

Transfer Stations

The expansion of cities during recent decades has created several problems in the siting of new final disposal sites. Large distances between the generation and final disposal sites make transportation costs high. At transfer stations, waste can be transferred from trucks with a capacity of 2-5 tons and a crew of one driver and two workers to bigger trucks with a capacity of 20-25 tons, and crew reduced to one driver, which reduces the unit cost per kilometre. Large metropolitan areas in the LAC region, such as Mexico, Buenos Aires, Rio de Janeiro and Bogota, send around 50% of all MSW to transfer stations.

Street Cleaning

Street sweeping is one of the SWM aspects in which public awareness and education play critical roles. In medium- and low-income areas, a considerable amount of street waste is generated by deficiencies in the refuse collection system (Diaz *et al.*, 1996). The cost of collecting litter from the streets is considerably higher than the cost of collecting waste in conventional containers. In most of the LAC cities, one worker sweeps 1-2 km of street per day (which means 2-4 km of pavement per day) and collects 30 to 90 kg of waste per km. For each 1,000 inhabitants 0.4-0.8 workers are required, depending on technical aspects, such as whether the streets are paved or not, the availability of mechanical sweeping devices, etc. In cities with beaches, the number of street cleaners is very high. The percentage of streets, which are swept in cities in the LAC region, varies from 10-100%.

Material Recovering and Recycling

Source separation of recyclable material, followed by special collection and transportation, has been successfully implemented in most developed countries, and is now a basic prerequisite for the recycling industry. However, this method of collection cannot be rapidly implemented on a large scale in LAC countries but is a process requiring time for education and adaptation. It has also been found that the collection of source-separated recyclables conducted by municipalities in medium-size and large cities in the LCA region was significantly more expensive than traditional collection methods. Unless changes in the system succeed in reducing the cost to an affordable level, source-separated collection may not be sustainable in a long-term perspective. Separation at source programmes conducted by municipalities in six Brazilian medium-size/large cities according to the investigation of CEMPES in 1995, demonstrated costs ten times higher than those for conventional collection: (US$ 262/ton versus US$ 26/ton). Nevertheless, the recycling industry in many LAC cities has been supplied with recyclable waste sorted and traded mainly by the unofficial sector, namely scavengers working at different points along the waste stream, and a wide range of intermediate traders working between scavengers and industries (Gomes and Hogland, 1995). Mechanical sorting plants for mixed waste employing technologies with different degrees of complexity, (usually associated with a composting area for the organic fraction) were set up in many cities in the LAC countries, but most of them were closed after a few years of operation. Apart from the low degree of recovery and the low quality of recyclable materials and compost, such plants represent a high cost for most municipalities. In Brazil, among 57 such facilities in 1990, 18 were active, 15 were under construction and 24 had been closed (data from the Institute of Technological Research, IPT in Acurio *et al.*, 1997). In 1997, some of the plants under construction in 1990 had been closed. Sorting plants which receive only recyclable material without organic fraction can easily recover 80% or more of the marketable grades of metal, glass, plastic and paper (Diaz *et al.*, 1996). Such plants exhibit better results than sorting plants that receive mixed waste. But again, the cost related to collection of source-separated waste must be borne in mind.

Treatment and Final Disposal

To date, landfilling has been the most common method of waste treatment all over the world (Table 6). In the future, however, according to the directives of the European Union, landfilling in Europe will be greatly restricted. In Sweden, for example, to make landfilling less attractive, a national tax of 25 US$/ton of all landfilled waste has been proposed to take effect in 1998. Simultaneously, incineration will be allowed to increase, without any external restrictions. In other developed regions (e.g., the U.S.A.) landfilling will probably continue longer as the most common method of waste treatment. Apart from a reduction in the total volume, the waste remaining to be landfilled in Europe in the future will be of a different composition. Alternative treatment options to landfilling (e.g., incineration) have been submitted to long-term testing, improvement and adjustment in developed countries. Technical improvements and adaptation have been followed by the development of economically sustainable systems, since some treatment options may be 20 times more expensive than landfilling. In the LAC region, as well as other economically developing regions, most alternative treatment plants such as incinerators, sorting and composting plants for MSW have failed due to economic; technical and environmental problems. The exception to this is a number of small-scale composting plants and facilities that recover biogas from landfills.

Table 6. Forms of treatment and final disposal in various developed countries and in the LAC region (data from OPS, 1995).

Country/region	Landfilling (%) (sanitary, controlled, uncontrolled)	Incineration (%)	Composting (%)
LAC region	98	< 1	< 1
Spain	80	15	5
USA	80	19	< 1
Germany	67	30	3
France	55	40	5
Sweden	40	55	5
Japan	28	70	2
Switzerland	20	80	< 1

In the LAC region, about 30% of the final disposal sites for MSW are sanitary landfills, 35% are controlled landfills and the remaining 35% are open dumps (Table 7).

This basically describes the treatment of waste in the region, since 98% of the waste collected is landfilled. Among LAC countries, Chile has achieved the highest percentage of waste disposed at sites with good technical standards and half the Chilean cities have sanitary landfills.

One of the environmental problems related to landfilling is the production of leachate. Large landfills receive 5,000-7,000 tons/day (e.g., in Rio de Janeiro, São Paulo and Mexico City) and produce leachate around 1000 m³/day. Some of these landfills cannot send their leachate to the municipal wastewater treatment plant, but have to run on-site treatment facilities. Since many landfills receive not only MSW but also industrial waste with a hazardous content, it is necessary to choose a suitable leachate treatment technique.

One of the operational problems identified at landfills in small cities (less than 50,000 inhabitants) in the LCA region is related to the relatively high investment costs for tractors and compactors. To solve this problem, some small cities in Colombia and Chile, for example, have succeeded in implementing manually operated landfills.

While European countries are implementing means of restricting

Table 7. Collection and final disposal of MSW in some cities/metropolitan regions in LAC countries, with 0.5 million inhabitants or more (based on Acurio et al., 1997).

City (year)	Inhabitants (million)	Waste generation (ton/day)	Collection (%)	Final disposal site[2] (%)			Institution responsible	Service operated by:	Revenue/Cost Ratio[3]	Number of employees	Employees per 1000 inhabitants	Tons per employee per day
				Good	Medium	Bad						
M.R.[1] São Paulo (96)	16.4	22100	95	100	0	0	municipality	private	good	10000	0.6	2.2
M.R. México (94)	15.6	18700	80	50	25	25	municipality	municipality	bad	17000	1.1	
M.R. Buenos Aires (96)	12.0	10500	91	100	0	0	Municipal company	private	good			
M.R. Rio de Janeiro (96)	9.9	9900	95	0	100	0	Municipal company	mixed	medium	12000	1.2	0.8
M.R. Lima (96)	7.5	4200	60	0	40	60	Municipal company	municipality	bad	5500	1.2	0.5
Bogotá (96)	5.6	4200	99	100	0	0	Municipal company	private		2600	0.5	1.6
Santiago (95)	5.3	4600	100	100	0	0	Municipal company	private	good			
Belo Horizonte (96)	3.9	3200	90	100	0	0	Municipal company	mixed				
Caracas (95)	3.0	3500	95	0	100	0	Municipal company	private	bad	5110	1.8	0.7
Salvador (96)	2.8	2800	93	0	100	0	Municipal company	mixed		2345	0.8	1.2
M.R. Monterrey (96)	2.8	3000	81	0	100	0	Municipal company	mixed				
Santo Domingo (94)	2.8	1700	65	0	0	100	municipality	mixed	bad			
Curitiba (95)	2.1	1300	100	100	0	0	municipality	private				
La Habana (91)	2.0	1400	100	0	100	0	municipality	private		1800	0.9	0.8
Guayaquil (96)	2.3	1400	100	100	0	0	municipality	mixed	good	843	0.4	1.6
Brasilia (96)	1.8	1600	95	0	75	25	Municipal company	private		745	0.4	2.1
Cali (96)	1.8	1350	95	0	0	100	Municipal company	mixed		1313	0.7	1.0
Medellin (87)	1.5	750	99	100	0	0	municipality	municipality	good	750	0.5	1.0
Montevideo (95)	1.4	1260	97	0	0	100	Municipal company	municipality	good			
Quito (94)	1.3	900	85	0	0	100	municipality	municipality	S/D	2443	1.7	0.5
Guatemala (92)	1.3	1200	80	0	0	100	municipality	municipality	bad	594	0.4	2.0
San Salvador (92)	1.3	700	60	0	0	100	municipality	municipality	good	1150	0.9	0.6
Asunción (96)	1.2	1100	80	0	0	0	Municipal company	mixed	medium	1100	0.9	1.0
Rosario (96)	1.1	700	100	100	0	0	municipality	private	medium			
San José (95)	1.0	960	90	100	0	0	municipality	municipality	good	900	0.9	1.0
Managua (88)	1.0	600	70	0	100	0	municipality	municipality	S/D			
Tegucigalpa (95)	1.0	650	75	0	0	100	municipality	municipality	medium	480	0.5	1.4
Bartranquilla (96)	1.0	900	98	0	0	100	municipality	mixed		659	0.7	1.4
La Paz (96)	0.7	380	92	100	0	0	municipal company	private	bad	450	0.6	0.7
Joao Pessoa (96)	0.7	250	95	0	60	40	municipality	mixed		730	1.0	0.3
Cartagena (96)	0.6	560	96	0	100	0	Municipality	mixed				
P. Spain (93)	0.5	600	98	0	100	0	Municipal company	mixed	bad			
Total	113.1	109130	89	30	35	35	Municipality 47%	Municipality 37%		71712	0.9	1.2

[1] M.R.=Metropolitan Region [2] Good: sanitary landfill, Medium: controlled landfill, Bad: open dump [3] Good: I/C > 0.99; Medium: I/C = 0.66-0.99; Bad: I/C < 0.66

landfilling to pre-treated (rest) waste, in LAC countries the main aims for the coming years are to construct as many sanitary landfills as necessary, preferably with energy recovery through biogas exploitation, and, at the same time, to remediate existing open dump sites.

FINANCIAL AND ECONOMIC ASPECTS

Operating Expenses

In the LAC region, the MSWM tax ranges from US$ 0-5 per person per month, while in the U.S.A. it ranges from US$ 20-30 per person per month. While revenues may cover primary collection costs, in the LAC cities they seldom cover full transfer, treatment and disposal costs, especially among low-income groups. Therefore, to achieve equity of waste service access, some cross-subsidization and financing out of general revenues have been proposed (Schübeler, 1996). However, large-scale generators should pay the full cost of disposal services based on the "polluter pays" principle.

The most common way of charging for MSWM services in LAC cities is inclusion in the urban property tax. The inefficiency in tax collection in many municipalities thus results in a poor financial status for the solid waste management sector. An itemized tax bill identifying the waste tax should be the rule, but this is seldom seen. The revenue flows into a general municipal account, where it tends to be absorbed by overall expenditure instead of being diverted to the intended purpose. Another way of charging for MSWM services is inclusion in the electricity or water tax. This system has worked well in some LAC countries where legislation does not proscribe such an option. A third alternative is to employ a totally independent charging system for MSWM services through user charges where the contractors which are responsible for the services, charge their customers directly without participation of the municipality. This system is more common in developed countries but it is not a welcome alternative among contractors in the LCA cities. To improve the operational efficiency and quality of the services and to give transparency regarding the real costs, proper systems of budgeting and cost accounting for MSWM should be urgently established in the LAC cities.

External Financial Support

During the two-year period 1996-1998, the Interamerican Development Bank (IDB) supported about 30 programmes concerned with sanitation, environment and urban development in the LAC region. SWM projects were included in these programmes were implemented in Argentina, the Bahamas, Barbados, Bolivia, Brazil, Colombia, Chile, Costa Rica, Ecuador,

El Salvador, Guatemala, Guyana, Jamaica, Mexico, Paraguay, Trinidad and Tobago and Uruguay. The total loan for these programmes was around US$ 3.3 × 10⁹ for a three-year period and of this, US$ 493 × 10⁶ (15%) were invested in SWM projects (IDB data, in Acurio *et al.*, 1997).

During the nine-year period 1988-1996 the World Bank supported 24 environmental/social programmes in the LAC region which included SWM, contributing with US$ 2.2 × 10⁹ from a total budget of US$ 5.4 × 10⁹. Of this budget, US$ 430 × 10⁶ (8%) were invested in MSWM projects (Table 8). During the period 1995-2000 the World Bank will support SWM projects in the Caribbean with a total loan of US$ 11.5 million to the following countries: Dominica, St. Kitts and Nevis, St. Lucia, San Vicente and Las Granadines (World Bank data, in Acurio *et al.*, 1997).

Table 8. Loan from the World Bank to LAC countries and associated national investments for environmental programmes: period 1988-1996.

Country	Number of projects	Total cost (US$ × 10⁶)	Loan from World Bank (US$ × 10⁶)	Other sources (National, etc.) (US$ × 10⁶)	MSWM part (US$ × 10⁶)	MSWM %
Antigua Barbuda	1	50.5	6.8	43.7	50.5	100
Argentina	2	840	330	510	23.6	2.8
Belize	1	28	20	8	0.5	1.8
Bolivia	1	21.3	15	6.3	2.2	10.3
Brazil	8	2693.1	1098.9	1594.2	97.2	3.6
Chile	1	32.8	11.5	21.3	0	0.0
Colombia	4	697.4	251.9	445.5	15.8	2.3
Ecuador	1	300	104	196	5	1.7
Mexico	3	617.5	285.8	331.7	223.5	36.2
Peru	1	52.5	24.7	27.8	6.6	12.6
Venezuela	1	85.5	40	45.5	4.7	5.5
Total	24	5418.6	2188.6	3230	429.6	
%		100	40.4	59.6	7.9	
Average per project		225.8	91.2	134.6	17.9	7.9

Source: World Bank data, in Acurio *et al.*, (1997).

Non-governmental Organization for MSWM in Low-income Areas

One interesting alternative conducted in LAC countries by local organizations, sometimes with external support concerns the development of micro-enterprises for waste management. In Peru, for example, there are more than 150 micro-enterprises in the waste sector. Micro-enterprises have provided a viable option for collection services in marginal and poor areas. In most of the cases, residents of the area served are requested to participate. Sometimes, the organisation starts locally as a cooperative that becomes a micro-enterprise. The German agency GTZ has successfully assisted the setting up of micro-enterprises in Bolivia, Costa Rica, Colombia and Ecuador. A number of institutional arrangements "community-based" implemented in developing countries including the LAC region have been described and evaluated (Pfammatter and Schertenleib, 1996).

Regarding training and education of personnel at the university level, the following institutions have been active in the LAC region: Panamerican Health Organization (PHO), the Japanese International Cooperation Agency (JICA), Panamerican Center of Sanitary Engineering (CEPIS/PHO), and a number of universities, particularly in Brazil, Argentina, Chile, Mexico and Colombia. Regarding intermediate and basic education for workers and administrative staff, there is a lack of training in most cities in the LAC region.

CONCLUSIONS

In LAC region, the collection and transportation of waste as well as street cleaning in central and high-income districts of medium and large cities have achieved a good standard concerning quality of services. The main goal in the near future is to improve the collection at marginal and low-income areas, to establish independent management systems for hazardous waste with regard to collection and transportation, as well as to improve treatment methods and final disposal sites for municipal, industrial and hazardous waste. To achieve this goal, an effective institutional framework for MSWM must be established and changes should be made in the charging and financial system in order to implement sustainable management models before promoting increase in operational costs with new facilities.

ACKNOWLEDGEMENT

The authors wish to thank the Swedish Board for Industrial and Technical Development (NUTEK), the Swedish International Cooperation Development Agency (SIDA) and the Swedish Institute (SI) which have given them financial support. The authors have been supplied with recent reports published by the World Health Organization through the Panamerica Health Organization, which have been valuable for their work.

END NOTE

[1] The term privatisation is used here in a broad sense, including different kinds of private sector participation, such as third-party services, concessions and the free market, the latter being common for industrial waste.

REFERENCES

Acurio, G., Rossin, A., Teixeira, P.F. and Zepeda, F. 1997. Diagnóstico de la situación del manejo de residuos sólidos municipales en América Latina y el Caribe. Ed. Banco Interamericano de Desarrollo y la Organización Panamericana de la Salud. Washington DC, 155 pp.

CEPAL. 1995. La evaluación de impacto ambiental y la gestión de residuos. Comisión Económica para América Latina y el Caribe (CEPAL). Santiago, 1995.

Diaz, L.F., Savage, G.M., Eggerth, L.L. and Golueke, C.G. 1996. Solid waste management for economically developing countries. Ed. ISWA. Cal Recovery Inc. ISBN 87-90402-01-4. California, 1996, 416 pp.

ECLAC. 1998. Preliminary Overview of the Economies of Latin America and the Caribbean: 1998 Report. Summary. Ed. Economic Commission for Latin America and The Caribbean (ECLAC). Santiago, 1998.

EUROSTAT. 1994. Europe's Environment 1993: Statistical Compendium. Statistical Office of the European Community, June 1994.

Gomes, M.M. 1995. O Componente Resíduos Sólidos no Projeto Reconstruão-Rio e no Programa de Despoluicão da Baia de Guanabara: Desafios e Licões. In: Saneamento Ambiental na Baixada: Cidadania e Gestão Democratica. Ed. FASE and Interamerican Foundation. Rio de Janeiro, 1995. pp. 107-124.

Gomes, M.M. and Hogland, W. 1995. Scavengers and landfilling in developing countries. Sardinia '95. *Proceedings of the Fifth International Landfill Symposium*, 2-6 October, 1995, Cagliari, Vol. I, pp. 875-880.

Organizacion Panamericana de la Salud (OPS). 1995. El manejo de residuos sólidos municipales en América Latina y el Caribe. Serie Ambiental N. 15. Washington, DC, 1995.

Pfammatter, R. and Schertenleib, R. 1996. Non-governmental refuse collection in low-income urban areas. Ed. SANDEC/EAWAG. Report N. 1/96. Switzerland, 1996.

Schübeler, P. 1996. Conceptual framework for municipal solid waste management in low-income countries. Urban management programme (UMP) paper series 9, Swiss Agency for Development and Cooperation (SDC), August 1996. 59 pp.

United States Environmental Protection Agency (US EPA). 1995. Characterization of municipal solid waste in the United States: 1995 Update. EPA, March 1996. 134 pp.

White, P.R., Franke, M. and Hindle, P. 1995. Integrated solid waste management: A life cycle inventory. Chapman and Hall. Blackie Academic & Professional. Glasgow, 1995. 362 pp.

22

Solid Waste as Ore:
Scavenging in Developing Countries

Marcia Marques[1,2] and William Hogland[1]

[1] Department of Technology, University of Kalmar,
PO Box 905, SE-391 29 Kalmar, Sweden

[2] Department of Environmental Engineering,
Rio de Janeiro State University, Rio de Janeiro, Brazil

INTRODUCTION

This paper deals with scavenging activities at landfills in Rio de Janeiro State, Brazil. This activity may be considered a specific form of "casual work" distinguishable from "stable wage work" characteristic of capitalism production.

In contrast to Blincow (1986), who classifies scavengers into four categories including all those involved in handling of waste products, we prefer Sicular's (1991) definition based on the labor process itself. He considers as scavengers only those who treat waste as ore: a source from which valuable materials can be extracted. Those who treat waste as waste are referred to as "refuse workers".

A further distinction is made between those who purchase materials from their owners before they enter the waste stream, in which case, the materials are seen by all parties as still having some value, and those who produce materials from what has already become waste. Following again Sicular's (1991) definition, only the second group is considered as scavengers. Based on scavenging activities in Bandung, Indonesia, Sicular (1991) considers scavenging as a form of non-capitalist production connected with the capitalist sector through the market. The specific form this production takes would be technically similar to hunting and gathering and so, structurally similar to Friedman's (1980) political-economic concepts of peasant production.

The existence of scavenging as a distinct occupation is based on the following conditions: (1) a market for recovered materials; (2) waste in sufficient quantity and quality to meet industrial demand; and (3) people who are willing or compelled to do work that is poorly paid, hazardous and of low status. These conditions are satisfied in developing countries such as Brazil which shows a good industrialization standard, increasing population with no qualified labor, and migration from smaller cities and rural areas to big cities.

In countries lacking a formal welfare system, scavenging is the ultimate safety net occupation. It is still largely open to anyone, and free for penetration by organized crime, as has happened elsewhere in the developing world. In cities where the solid waste management is not good, scavengers benefit from open access to waste bins mainly in urban areas and to the open dumps around these areas.

Based on World Bank estimates, Soerjani (1984) pointed out that 1-2% of the population of big cities is supported directly or indirectly by the refuse generated by the upper 10-20% of the population. Scavenging as a whole not only provides a source of income for one of the poorest segments of the population, but also reduces the need for highly sophisticated and costly recovery systems. Although there are several socioeconomic and hygienic problems associated with scavenging, the scavengers nevertheless perform an important function in the recovery and recycling of materials.

Although the elimination of scavenging is desirable from the humanitarian point of view, it nevertheless constitutes an important form of solid waste management. Thus, any attempt to improve solid waste management in a city, or to promote the environmental recovery of landfill sites should only be made after implementing alternative means of survival for scavenger groups. The simple banning of scavengers from a site with the purpose of environmental recovery would force them to move either way along the collection system: towards the waste source or towards the waste destination. In Pakistan, it was demonstrated that where collection at the doorway is relatively efficient, and piles of waste in commercial areas are discouraged, scavengers that used to work among the piles of commercial waste now go through the plastic bags put out by households. It was pointed out that this sorting close to the source is just one step way from source separation.

There is a difference in status among the groups of scavengers, based on the scavenging site. Due to the symbolism associated with waste, the scavengers working in landfills and open dumps are considered, in most cultures, to have the lowest social status.

During the last decade, associations and self-reliant community institutions in many cities in developing countries have been active in supporting the efforts of the scavengers. However, in spite of their efforts, most of them have not succeeded in improving the scavengers' quality of life. This is probably related to the dependency of the scavengers on their "buyers" or receivers. This must be eliminated before a true transformation of social status can be achieved. This means that any waste management programme in developing countries must include scavenging as an important aspect, to be understood and considered from the beginning. The identification of these groups, their labor process and social aspects related to production are important for any action in this complex system.

SCAVENGERS' PROGRAMME IN THE STATE OF RIO de JANEIRO

The Government of Rio de Janeiro State started the Guanabara Bay Clean-up Programme in March 1994, with a loan from The Interamerican Development Bank (IDB) and a Japanese Fund, The Overseas Economic Co-operation Fund (OECF).The Programme included activities in different areas such as: sewerage, water supply, drainage, solid waste management, industrial pollution control, environmental education, digital mapping and tax levying.

Solid waste management projects for seven municipalities in the basin of Guanabara Bay include the following activities:
- recovery of two old landfills, transforming them into sanitary landfills;
- construction of a new sanitary landfill;
- construction of three sorting and composting plants;
- installation of five incinerator plants for medical waste;
- a new administrative approach with emphasis on training and third part services;
- equipment.

As an important part of the two old landfill recovery projects for the Municipalities of Niteroi (Morro do Céu landfill) and São Gonçalo (Itaoca landfill), a social programme was carried out with around 300 scavengers who had been working there for several years. It was realized that no environmental recovery of these old landfills could be conducted while maintaining the scavengers' activities. On the other hand, the scavengers could not just be driven off the sites, due to the social problems this would cause. The social programme for scavengers was created to face the challenge of transforming open dumps into sanitary landfills, at the same time giving the scavengers the possibility of a better quality of life. The alternatives to scavenging had to be compatible with their occupational, social, and educational profiles.

An NGO called the Federation of Social and Educational Assistance-FASE/RJ was contracted to conduct this programme together with the State Government and the Municipal governments of Niteroi and

São Gonçalo. The programme was to precede the recovery of the open dump sites. It was divided into three stages: (i) social group characterization; (ii) analysis of the alternatives, and (iii) implementation.

This paper present paper presents the results of the first stage for the Itaoca landfill. The results are compared with those obtained in a similar study conducted by the Municipal Company of Public Cleansing of Rio de Janeiro city, through Engevix (1993) at the Metropolitan landfill where 570 scavengers are still working.

Both Itaoca and Metropolitan landfills had in common, serious problems concerning environmental damage, social problems and poor maintenance, in spite of the fact that different sectors were responsible for their operation: a private company at Itaoca landfill and a municipal company in the Metropolitan landfill. This example illustrates a simple principle: in developing countries, problems related to the quality of sanitation services are more complex than the discussion concerning comparative efficiency of the private sector versus the governmental sector.

CHARACTERIZATION OF THE SOCIAL GROUP

The objective of this stage was to have the identification and characterization of the scavenging community by the project team and to give the scavengers the opportunity to get to know the members of the project team, in order to reduce anxiety due to the announced changes. The scavengers initially shunned all contact with the project team. In fact, they reacted against foreigners, in general due to previous experiences when visitors emphasized only the low social status of scavenging.

The strategy for this stage also included the development of the awareness of the local authorities, the neighbourhood and the scavengers themselves about the social role of the scavenging and the environmental benefits of recycling.

During this period, the project team visited the landfill many times in order to establish good relationships with the group. Identification documents were ordered for scavengers who lacked them, and social activities were organised on site, with the participation of the scavengers as players and singers. An exhibition of photographs was organized at the entrance of the open dump. The project team contacted the municipal secretaries of health care, education and social development with the aim of initiating specific actions such as medical care and development of a educational course for illiterates.

Through on site observations and a questionnaire it was possible to identify some important aspects of the labor process and the production. These are described as follows:

ITAOCA LANDFILL CHARACTERISTICS

This landfill occupied an area of 230,000 m². The distance between the landfill and the city centre was 8 km. The landfill was 4 km from an important road that crossed the municipality. It was close to a mangrove area classified as an environmental protection area. According to the local community, this site had been used as an open dump for the past 20 years. A weighbridge was installed only three years earlier and before this, there was no control of the amount of garbage that was disposed of. The principal economic activity surrounding this site was the recovery and trade of recyclable goods as a consequence of the long time this site had been used for waste disposal and scavenging. There were also some small shops in this area which accepted recyclable goods as money to buy goods, mainly food.

About 100 trucks came daily to this landfill, which was operated by a private company contracted by the local government. Although there had been many complaints during the preceding decade concerning the environmental and social problems of this site, the State Environmental Agency had not succeeded in closing the landfill. The local government argued that there was no proper place to put the waste, and no financial support to recover the area and construct a new landfill. The scavengers and dealers were aware of the contention that had prevailed during the previous decade. They have heard about projects that had never come to fruition, and therefore refused to discuss changes. As a consequence of this, the first problem the project team faced when they started the programme was to convince the scavengers about the necessity of discussing a strategy for the future, when scavenging would not be allowed any longer.

SOCIAL ASPECTS

The interviews at the Itaoca landfill included 112 scavengers. Only scavengers older than 14 years were included in this study. To start working with the children it was necessary, at first, to start working with their parents and convince them of the importance of sending the children back to school. The parents would need give up the income generated by the children's scavenging.

At Itaoca landfill there was a balance between the different age groups (Fig. 1). The groups were: 14 to 17 years old (11%), 18 to 30 years old (25%), 31 to 45 years old (35%), 46 to 65 years old (27%), older than 65 years old (3%). Most of the scavengers had relatives working on the landfill, sometimes, the whole family. The sex ratio also was balanced (54% men).

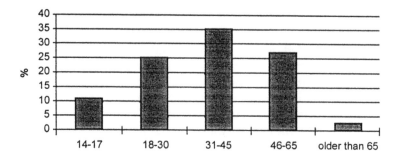

Fig. 1. Age distribution. Scavengers at Itaoca landfill

The educational level, considering only reading and writing ability was 38% illiterate and 55% literate. In the second group, only 4% had been to school after primary school (Fig. 2). There were no illiterates among the teenagers and the concentration of illiterates was in the older group.

The probability of these people getting formal jobs was very low. Scavenging provided an income as it was not necessary to have any special ability but the ability to sort out recyclable goods, the income being determined by the amount and type of recyclable goods sorted.

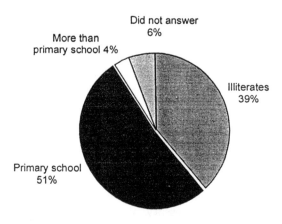

Fig. 2. Education level. Scavengers at Itaoca landfill

The most frequent activities of scavengers when they were not working at the landfill were: cleaning home, taking part in parties (playing, dancing and singing), playing football, going to the beach, listening to the radio, or watching TV. Only 3% of the scavengers pursued any regular religious activities.

LABOR PROCESS

Most scavengers went to work by foot (89%). The others hitchhiked (10%), or took a bus (1%). Most of them spent 10-30 minutes to reach the dumpsite. The lack of or low transportation costs was an economic incentive to work at the landfill. Most of the scavengers (77%) worked eight hours or more daily. In fact, they needed to remain at the dump site during many hours each day, even if they are not scavenging. To get enough recyclable goods from the waste it is necessary to wait for the trucks with the "good waste". For the same reason, they mostly worked seven days per week (16%), or six days per week (35%), or five days per week (38%). Only a small group (8%) works four days per week and 3% did not answer (Figs. 3 and 4).

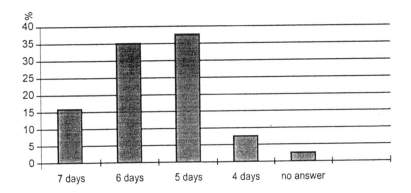

Fig. 3. Working days per week. Scavengers at Itaoca landfill

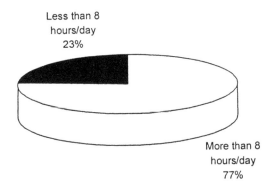

Fig. 4. Working hours per day. Scavengers at Itaoca landfill

The long periods spent at the same site gave them the opportunity to establish strong social contact with others, which could either result in friendship or conflict.

Other informal workers also worked at the landfill (9% of the whole group). They did not scavenge but sold beverages or kept pigs at the landfill which ate food waste found in the landfill.

An interesting observation is that most of scavengers at Itaoca landfill sort all kind of recyclable goods (89%) such as: aluminium cans, glass, paper, cardboard and plastic and only a small group (11%) looks for a specific material. This will be considered later in a comparison with Metropolitan landfill.

WORKING BACKGROUND

Most of the scavengers (86%) had a formal job in the past. Their working backgrounds varied considerably, but the most frequent professions mentioned were: maids (38%, women only), civil servants (27%, men only), metal industry workers (8%, men only) and commerce workers (7%). In contrast to other groups of scavenger described in developing countries, only 2% of them came directly from rural areas to scavenge. The main reasons why they left their jobs were (Fig. 5): dismissal (27%); salaries too low (21%); need to take care of the children (15%, women only); intention of working by themselves (12%). The working backgrounds of the scavengers and the reasons why they left their jobs confirms that such situation were a consequence of national and local economic changes which severely affected the poorest.

INCOME

Different kinds of waste have different market values which may change during the year. The income of the scavengers therefore, depended on their

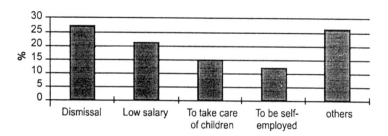

Fig. 5. Main reason for leaving previous jobs. Scavengers at Itaoca landfill

ability to find valuable materials, or to sort faster in a competitive way. It also depended on the amount of waste they were able to sort out. Incomes therefore, varied considerably: from less than US$ 88 to more than US$ 224 per month. Considering that the minimum Brazilian salary was around US$ 100 at the time, this explained why most of them considered waste to be "good business" (66%).

An interesting result arose when the women's and men's opinions were analysed separately: while 85% of men considered scavenging to be good business, only 52% of women were of the same opinion. This difference was probably due to the lower income for women who usually lost in the competition for more and better recyclable material. These results contradict the higher women's productivity observed at some sorting plants in Brazil. Probably, because the intimidation by physical supremacy observed at landfills is not upheld at the conveyor belts in sorting plants.

EXPECTATIONS

The advantages of working at the landfill, according to the scavengers, were (Fig. 6): no boss (23%); better income than that obtained through formal jobs (13%); goods for personal use found in the waste (10%); find food to eat (9%); food to feed animals (5%); none (3%); others (7%); don't know (30%). The disadvantages were (Fig. 7): the bad working environment (21%); no formal contract (15%); accidents at the landfill (7%); none (6%); disease (5%); scavenging by the trucks' crews or other refuse workers before waste being finally deposited at the landfill (4%); low salaries (3%); others (10); don't know (30%).

According to 70% of the scavengers the working conditions could be improved, 22% did not believe that this was possible and 8% has no opinion at all.

Among those who believed that the working conditions could be improved, the main suggestions were: organize the scavenging in a better manner (41%); construct a recycling and composting plant (14%); transform scavenging into a formal job (5%); avoid scavenging by the trucks' crews (11%). The last suggestion resulted from the fact that by the time the waste arrived at the landfill it had ofteb already been picked over several times by itinerant scavengers and the waste collection crew who sold the sorted material to dealers before the balance was deposited at the final disposal site. This competition with the dump scavengers was one of the constant complaints of the scavengers and this is common in other countries as well (Keyes, 1982 and Fernandez and de la Torre, 1986 in Bubel, 1990).

When asked if they would like to have another job instead of scavenging, 70% said yes and 30% said that they preferred to continue scavenging.

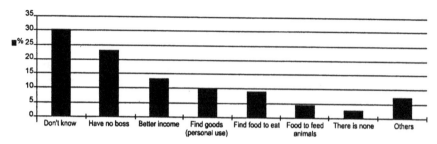

Fig. 6. Advantage of working as scavenger, according to scavengers at Itaoca landfill

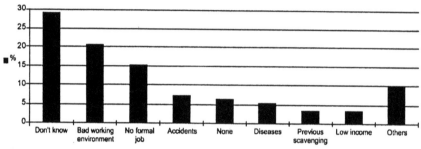

Fig. 7. Disadvantages of working as scavenger, according to scavengers at Itaoca landfill

Among those who would have liked another job, their wishes seemed to be related to their working background: maid (16%, women only); civil construction (11%); driver (10%); commerce worker (10%); mechanic (9%) and night watchman (9%).

SCAVENGING AT THE METROPOLITAN LANDFILL IN RIO de JANEIRO STATE

This landfill covered 1.2 million m² and received in 1995 around 6,000 tons of waste per day from four municipalities, including Rio de Janeiro. At that time the Rio de Janeiro Municipal Company of Public Cleansing-COMLURB was trying to get some financial support to recover the area, stop the environmental degradation caused by the waste disposal and solve the social problems due to the growing scavenging. A reclamation project for the site started up in 1997. The community with 570 scavengers working at the Metropolitan landfill were identified and characterized in 1993 (Engevix, 1993). The results of this research were compared to those obtained at Itaoca landfill. The main results suggest some social aspects which need to be considered for implementing change. The age distribution showed a greater concentration of young people, from 18 to 30 years old in the metropolitan landfill community (43%), compared with the Itaoca landfill community (23%). While 77% of the scavengers at the Metropolitan landfill were men, at the Itaoca landfill the sex ratio was more balanced with 54% men (Fig. 8).

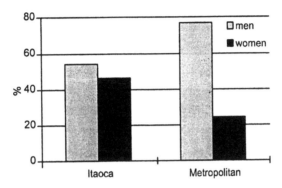

Fig. 8. Sex ratio at two landfills

Compared to the Itaoca landfill, most of the scavengers in the metropolitan landfill sorted the same material (77%) and only a smaller group (23%) changed the type of waste, mainly according to availability (Fig. 9).

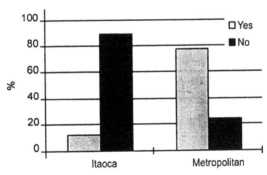

Fig. 9. Specialization in sorting at two landfills

Some results of the metropolitan landfill research indicate a relationship between age and productivity. The average productivity was 268 kg per scavenger per day. The group of 40 to 49 years old showed the highest productivity (334 kg per day per scavenger) while the group younger than 18 years old showed the lowest productivity (206 kg per scavenger per day). However, the income was higher in the groups of 18 to 29 and 30 to 39 years old.

The explanation of the positive correlation between age and productivity and between strength and better income is probably that the fastest and strongest group (18-39 years) dominated the sorting activity and picked up the most valuable material. They had a better income even if they worked less than the others. The oldest group had to deal with larger amounts of less valuable recyclable waste in order to get a reasonable income.

The youngest group (younger than 18) had the lowest productivity and also a lower income. Probably they could not compete with the strongest group of men and they did not work as hard or as long as the oldest scavengers.

CONCLUSIONS

The significant role played by the informal sector in recycling in developing countries with no effective environmental protection policy leads to the conclusion that recycling in these countries is a consequence of economic necessity rather than environmental policy. When comparing the answers to a questionnaire of scavengers' community working at Itaoca landfill, in Rio de Janeiro State, some contradictions arose. Although many scavengers considered waste to be good business, and mention the lack of a boss as an advantage, most of them would prefer to have a formal job. Scavenging gives them a feeling of freedom because they are not aware of the relation of dependence with the receivers. However, because of their awareness of the unhealthy working environment, scavengers want changes aimed at a higher degree of organization. Two variables influence the number of scavengers at a landfill: (i) the formal jobs available at a specific time, mainly for non-qualified labor; (ii) the amount of solid waste available at the site. Both variables regulate the income as well the size of these communities which change during the year.

Specialization towards scavenging a specific type of waste was conspicuous in the bigger community, but not in the smaller one, as was evident by comparing scavengers at Itaoca landfill with scavengers at the Metropolitan landfill. This was probably due to the competition which was stronger and the leadership of criminal elements at the bigger landfill which was also noted by the age of scavengers (younger at the bigger landfill). Under such conditions, only the strongest and youngest scavengers managed to set the valuable waste. The specialiszation of the others could be seen as a consequence.

To promote a real transformation in the quality of life of scavengers, it was important to first identify the dependency between the scavengers and the receivers. The information available was not sufficient to achieve a complete understanding of this relation. Deeper understanding is possible by studying the market structure for recyclable goods and the economic characteristics of the main participants in this sector. This includes dealers, wholesalers, small and medium-sized secondary-material industries, as well the virgin-material industries that act as competitors in this market. However, based on the available data, some strategies can be implemented in order to promote improvements in the scavengers quality of life. The methods used to reach this goal needs to take into account that scavenging a non-capitalist activity articulated with the capitalist sector through the market. To transform scavenging, any programme should introduce commodity relations and rationalization of production of recovered materials. Improvements may be made through social, technical or managerial assistance and credit, to transform the scavengers into an organized cooperative and, in a second stage, perhaps into a micro-enterprise. Another strategy is to give them formal jobs to operate recycling and composting plants under the administration of the local government. Based on the results presented in this paper, the first strategy is likely to be more promising.

ACKNOWLEDGEMENTS

The authors wish to thank the scavenger communities that took part in this project. The research activities in Brazil were financed by The Government of Rio de Janeiro State. Travelling costs for both authors were financed by The Swedish Institute and The International Co-operation Development Agency (Sida). A scholarship for the first author and the salary of the second author were financed by The Swedish Institute (SI) and The Swedish Board for Industrial and Technical Development (NUTEK), respectively.

REFERENCES

Blincow, M. (1986). Scavengers and recycling: a neglected domain of production. *Labor, Capital and Society*, 19 (1): 94-116.
Bubel, A.Z. (1990). Waste picking and solid waste management. *Environmental Sanitation Reviews*, December 1990, 30: 53-67.
Engevix Estudos e Projetos de Engenharia S.A. (1993). Projeto de Assimilação dos Catadores do Aterro Metropolitano de Gramacho: Relatório Final. Rio de Janeiro, Companhia Municipal de Limpeza Pública do Rio de Janeiro. *COMLURB*, 60 pp.
Friedmann, H. (1980). Household production and the national economy: concepts for the analysis of agrarian formation. *The Journal of Peasant Studies*, 7 (2): 158-184.
Gomes, M.M. and Hogland, W.H. (1995). Scavengers and landfilling in developing countries. *Proceedings Sardinia' 95. Fifth International Landfill Symposium*, CISA, Cagliari. I: 875-880.
Sicular, D.T. (1991). Pockets of peasants in Indonesian cities: the case of scavengers. *World Development*, 19 (2/3):137-161.
Soerjani, M. (1984). Present waste management in cities in Indonesia. *Conservation & Recycling*, 7 (2-4): 141-148

Milton Keynes UK
Ingram Content Group UK Ltd.
UKHW031127141024
449569UK00006B/377